Oxadiazole in Material and Medicinal Chemistry

Priya Ranjan Sahoo
Department of Chemistry
State University of New York at Buffalo
New York, United States

Abhishek Saxena
Department of Chemical Engineering
Hanyang University
Seoul, South Korea

Satish Kumar
Department of Chemistry
St. Stephen's College, University of Delhi
Delhi, India

CRC Press
Taylor & Francis Group
Boca Raton London New York

CRC Press is an imprint of the
Taylor & Francis Group, an **informa** business
A SCIENCE PUBLISHERS BOOK

Cover credit: Image used on the cover is prepared by the irst author, Dr. Priya Ranjan Sahoo.

First edition published 2025
by CRC Press
2385 NW Executive Center Drive, Suite 320, Boca Raton FL 33431

and by CRC Press
4 Park Square, Milton Park, Abingdon, Oxon, OX14 4RN

CRC Press is an imprint of Taylor & Francis Group, LLC

Library of Congress Cataloging-in-Publication Data (applied for)

ISBN: 978-1-032-47066-5 (hbk)
ISBN: 978-1-032-47067-2 (pbk)
ISBN: 978-1-003-38444-1 (ebk)

DOI: 10.1201/9781003384441

Typeset in Times New Roman
by Prime Publishing Services

Preface

Oxadiazole, a five-membered heterocyclic molecule affords multifaceted electronic utility. As a favorable acceptor, its quantum efficiency can be tuned with functional group engineering. As a result, many of us are using electronic gadgets every day, either in the form of laptops, smartphones liquid-crystal-display television sets, etc. It would be interesting to know the basic materials used in such devices during their assembly and manufacturing. This book explores various ways by which a variety of oxadiazole derivatives including its isomers could be synthesized. Further, it also discusses its significance and practical applications in electrochromic materials including electrochemical reduction, color change, and oxidation status.

Oxadiazole motifs are also widely found in bioactive drugs and medicinal compounds. It is one of the most treasured scaffolds for drug discovery scientists and biologists. Its aromatic flat surface facilitates biological target binding through π-stacking interactions. In the following chapters, readers will engage with oxadiazole-based inhibitors, cell imaging probes, and medicinally relevant molecules. The book is designed to promote small molecule heterocyclic scaffold in its prodigious applications in the areas of light-emitting diodes, liquid crystals, and electroluminescent features. It will provide a comprehensive account of various oxadiazole-based molecules where the reader will find material perspective in the initial section and the medicinal aspects in the later section. The detailed synthesis and practice would serve as a combo for the students and researchers practicing both material and medicinal domains.

Contents

Chapter 1

Diverse Synthetic Approaches to Design and Develop Oxadiazole-Based Heterocyclic Scaffolds

1. Introduction

The heterocyclic oxadiazole motif exists mostly in four different regioisomeric forms (attachment at different positions) (Figure 1). However, 1,2,4- and 1,3,4-oxadiazole motifs are most common in biologically active compounds (Boström et al. 2012). The 1,2,4-oxadiazole exhibits higher lipophilicity (log *D*) in comparison to 1,3,4-oxadiazole. Therefore, both these regioisomers display distinct difference in terms of hydrogen bonding potential, dipole moment, metabolic stability, etc. Tiemann and Krüger, in the year 1884, achieved the first synthesis of 1,2,4-oxadiazole (Tiemann and Krüger 1884). Since then, several ways have been explored to construct 1,2,4-oxadiazole moiety-based compounds. It can be constructed from nitriles via a 1,3-dipolar cycloaddition reaction to form nitrile oxide and 1,2,4-oxadiazole species (Pace and Pierro 2009). Alternatively, nitriles also react with hydroxylamine and activated carboxylic acid to form a 1,2,4-oxadiazole derivative. Additionally, *O*-acylation of amidoxime in the presence of an anhydride or acid chloride and subsequent cyclization often generate an oxadiazole scaffold. Similarly, substituted carboxylic acids and their conversion to symmetrical anhydrides in contact with EDC as a coupling reagent can also generate oxadiazole.

Oxadiazole derivatives possess inherent thermal as well as hydrolytic stability. The 1,2,4-oxadiazole moiety exists in a variety of medicinal compounds as pharmacophores, such as dopamine ligands (Carroll et al. 1993), benzodiazepine receptor agonists (Watjen et al. 2002), muscarinic agonists (Orlek et al. 1991), serotoninergic antagonists (5-HT_3) (Swain et al. 1991). Similar to 1,2,4-oxadiazole regioisomer, 1,3,4-oxadiazole has also been prepared via different routes including through oxidative cyclization (Gómez-Saiz et al. 2002; Jagadeesh Prathap et al. 2014; Niu et al. 2015), dehydrative cyclization (Yang et al. 2014; Tokumaru and

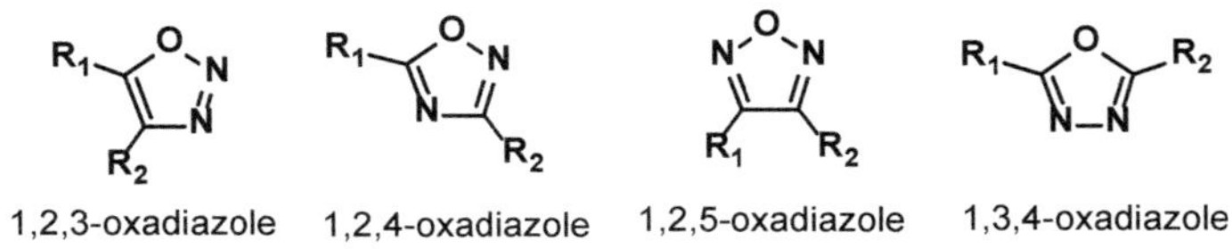

Figure 1. Structures of oxadiazole regioisomers.

Johnston 2017), Huisgen reaction (Mansoori et al. 2012; Wang et al. 2019), etc. Most routinely used protocols for oxadiazole ring constructions are discussed as follows.

2. Synthesis of 1,2,4-Oxadiazole Derivatives

2.1 Coupling Agent Assisted 1,2,4-Oxadiazole Synthesis

Carbonyldiimidazole (CDI) is an organic reagent useful for synthesizing various amides, peptides, and other organic scaffolds by activating carboxylic acid groups. One such application of CDI is to prepare oxadiazole-based heterocyclic compounds (Nandeesh et al. 2016). Deegan et al. synthesized oxadiazole derivative 1 (Scheme 1) from commercially available 4-methoxybenzoic acid in 97% yield (Deegan et al. 1999). The reaction of O-acyl benzamidoxime with CDI resulted in simpler isolation of oxadiazole compounds through liquid-liquid extraction followed by vacuum filtration.

Employing peptide coupling reagents in the reaction flask containing amidoxime and substituted carboxylic acid also leads to oxadiazole formation (Shaikh et al. 2023). N-(3-Dimethylaminopropyl)-N'-ethylcarbodiimide (EDC) is a water-soluble compound popularly known for its role in activating carboxylic acid. Liang et al. synthesized various 1,2,4-oxadiazole derivatives 2a–c (Scheme 2) using EDC as a coupling reagent (Liang and Feng 1996).

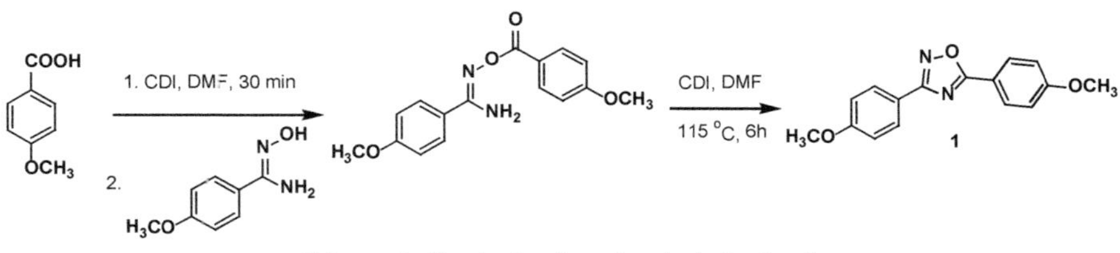

Scheme 1. Synthesis of oxadiazole derivative 1.

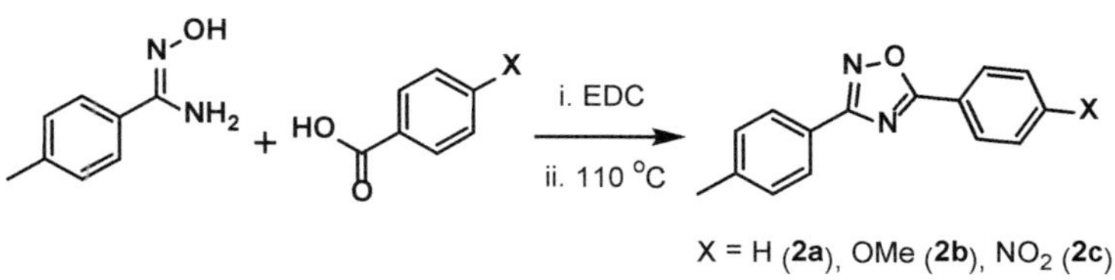

Scheme 2. Synthesis of oxadiazole derivative 2.

2.2 Pd Catalyzed Oxadiazole Synthesis

In 2014, ataluren (medication for Duchenne muscular dystrophy) was approved by the European Medicines Agency (EMA). Andersen et al. utilized a palladium catalyst to synthesize ataluren drug 3b (Andersen et al. 2014). The reaction employed 3-cyanobenzoic acid as the starting material, where the carboxylic acid was protected in the presence of boc anhydride (Scheme 3), and the terminal cyano group reacted in the presence of hydroxylamine to yield the corresponding amidoxime in 94% synthetic yield. The reaction facilitated the coupling of aryl bromide with amidoxime, resulting in 1,2,4-oxadiazoles in excellent yields. The protected tert-butyl group (in the case of 3a) was removed using trifluoroacetic acid (TFA) in dichloromethane.

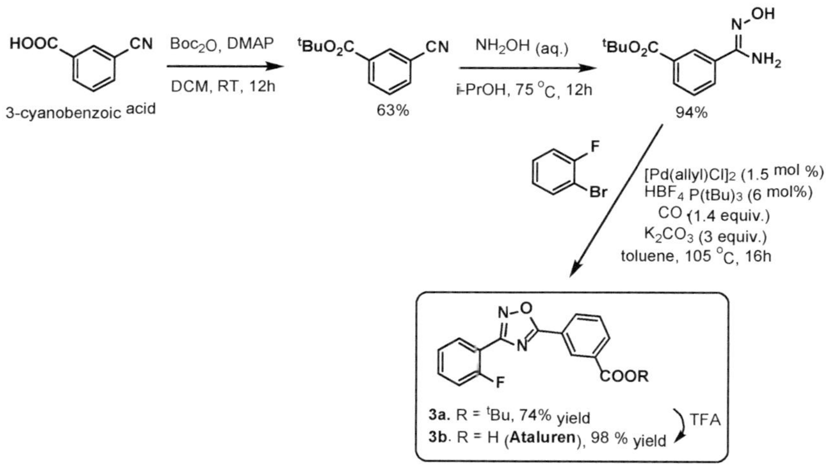

Scheme 3. Synthesis of oxadiazole derivative 3a–b.

2.3 Solid Phase-Based Synthesis of 1,2,4-Oxadiazole Derivative

Oxadiazoles have also been prepared using solid support (Sams and Lau 1999; Makara et al. 2002; Hu et al. 2008). The chemical reaction through solid-phase synthesis provides high efficiency as well as increased speed of the reaction. Scientists at Trega Biosciences (United States) led by Hébert et al. developed a series of oxadiazole derivatives on solid support starting from aliphatic and aromatic nitriles (Hébert et al. 1999). The authors utilized methylbenzhydryl amine (MBHA) resin during synthesis. The reaction of resin-bound nitriles in the presence of hydroxylamine hydrochloride (Scheme 4) and N, N-Diisopropylethylamine produced amide oxime intermediate. Subsequent reaction with a boc-protected amino acid anhydride and heating in the presence of 2-methoxyethyl ether yielded oxadiazole 4 in 95% yield. The purity of the synthesized product was subjected to exact synthetic protocol. Acylation of amide oxime required 60°C. The remaining amount of unreacted amino acid anhydride washed away, resulting in a robust synthetic tool.

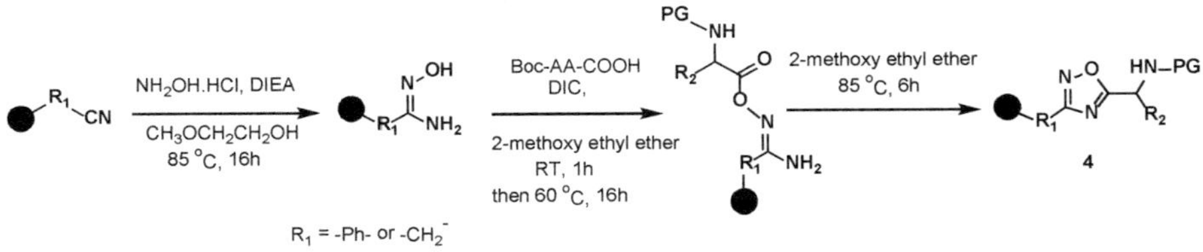

Scheme 4. Synthesis of oxadiazole derivative 4.

2.4 1,2,4-Oxadiazole Synthesis Using Oxidants

2.4.1 DDQ as Oxidant

Rapid and straightforward preparation of 1,2,4-oxadiazole has also been made possible by introducing oxidants such as 2,3-dichloro-5,6-dicyano-1,4-benzoquinone (DDQ). Parker and Pierce at North Carolina State University utilized DDQ to synthesize oxadiazole scaffold 5 from the corresponding amidoxime (Parker and Pierce 2016). A diverse set of 1,2,4-oxadiazoles was prepared from a range of aryl, alkyl, and heteroaryl substrates. The formation of oxadiazole involved two oxidation steps where one equivalent of DDQ promoted oxadiazole synthesis in 34% yield (Scheme 5). The introduction of another equivalent of DDQ increased the synthetic yield of oxadiazole from 34% to 48%. However, utilization of more than 2 equivalents of DDQ decreased the yield of reaction, indicating 2 equivalents of DDQ as the optimized condition.

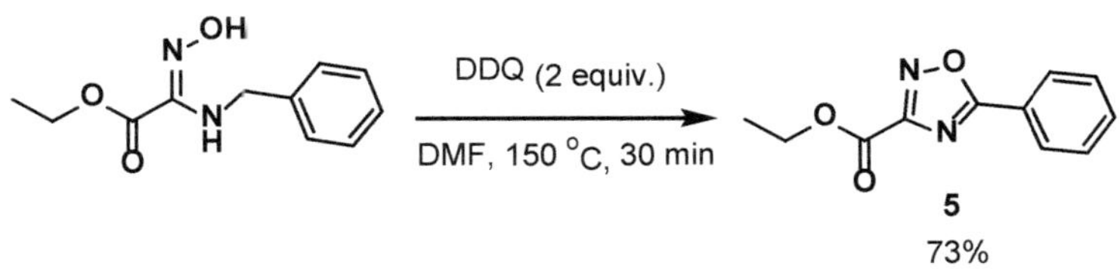

Scheme 5. Synthesis of oxadiazole derivative 5.

2.4.2 O_2 as Oxidants

Aerobic synthesis of 1,2,4-oxadiazole has been achieved by Zhang et al. in an O_2 atmosphere using N-benzyl amidoxime in 2013 (Zhang et al. 2013). Kuram et al. isolated oxadiazole 6 in 92% yield utilizing CuI and O_2 (Schemes 6–7) through an oxidative NO bond formation method (Kuram et al. 2016). This single-step synthetic process could transform amides and organic nitriles to their respective 1,2,4-oxadiazole scaffolds. Reaction yield increased to 22% in the presence of a base such as K_2CO_3. Similarly, the introduction of various nitrogen ligands was checked, and it was found that bathophenanthroline could increase the yield of the reaction up to 36%. Very interestingly, the addition of ZnI_2 increased the yield of the reaction up to 92%. Aromatic nitriles were found to be more reactive in comparison to their aliphatic counterpart.

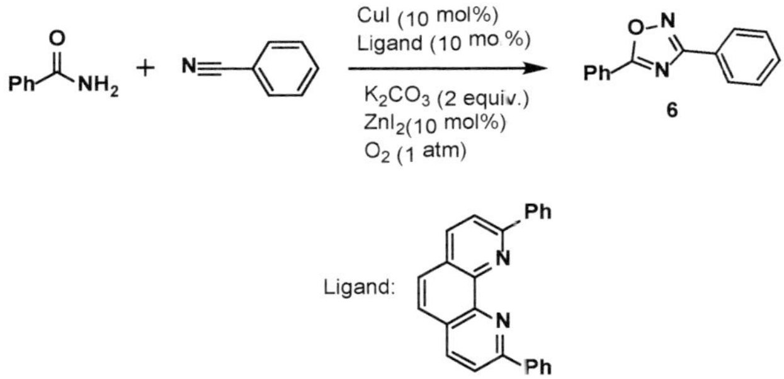

Scheme 6. Synthesis of oxadiazole derivative 6.

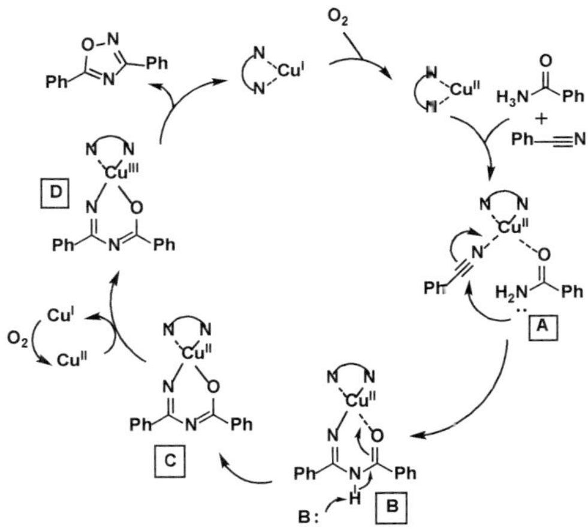

Scheme 7. Mechanism for the synthesis of oxadiazole (Kuram et al. 2016).

2.5 Visible Light Promoted 1,2,4-Oxadiazole Synthesis

Cai et al. developed a method for synthesizing 1,2,4-oxadiazole derivative 7 facilitated by visible light (Cai et al. 2019). In this method, substituted *2H*-azirines could react with nitrosoarenes and produce 1,2,4-oxadiazole through [3+2] cycloaddition reaction (Scheme 8). Nitrosoarenes acted as appropriate radical

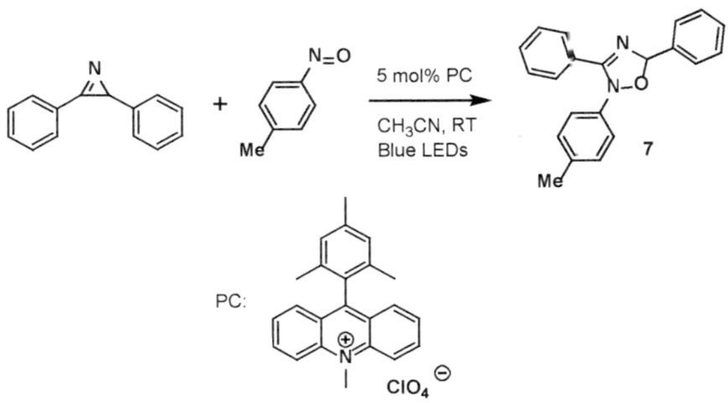

Scheme 8. Synthesis of oxadiazole derivative 7.

acceptors during the reaction. Also, 9-Mesityl-10-methylacridinium perchlorate is a well-known photo redox catalyst (PC) that promotes green synthesis of various chemical transformations assisted with visible light.

3. Synthesis of 1,3,4-Oxadiazole Derivatives

3.1 $POCl_3$-Promoted Oxadiazole Synthesis

Phosphoryl chloride promotes cyclization of reactions apart from its role in chlorination. Wang et al. synthesized oxadiazole derivative 8 in four synthetic steps utilizing pyridine-2,5-dicarboxylic acid (Wang et al. 2001). This synthetic process involved the methylation of carboxylic acid followed by a reaction with hydrazine hydrate (Scheme 9) to produce the corresponding hydrazide. It further reacted with 4-tertbutyl benzoyl chloride, subsequently reacted with phosphoryl chloride, and neutralized with concentrated NaOH solution to produce the oxadiazole compound 8 in 53% yield.

To find suitable bioisostere, Mohammed et al. synthesized oxadiazole derivative 9 as part of optimization studies (Mohammed et al. 2016). 2-Iodobenzoic acid acted as the starting precursor, reacting with thionyl chloride (Scheme 10) and hydrazine hydrate to produce the corresponding hydrazide intermediate. Further, the hydrazide intermediate reacted with 2-methyoxy benzoic acid in the presence of $POCl_3$ and produced the oxadiazole intermediate. Finally, it reacted with 4-nitrothiophenol in the presence of potassium carbonate and yielded compound 9.

In 2009, researchers at Syngene International Limited utilized propylphosphonic anhydride (T3P) to synthesize 1,3,4-oxadiazole derivatives (Augustine et al. 2009). This one-pot synthetic step involved the reaction of various carboxylic acid precursors with aryl hydrazides. Substituted acid hydrazide in the presence of CS_2 (carbon disulfide) and potassium hydroxide engaged in heterocyclic ring-closing and forms an oxadiazole scaffold from an ethanolic solution (Kaplancikli 2011).

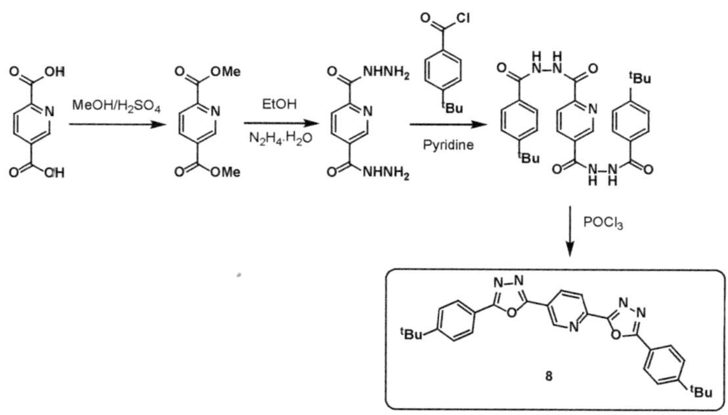

Scheme 9. Synthesis of oxadiazole derivative 8.

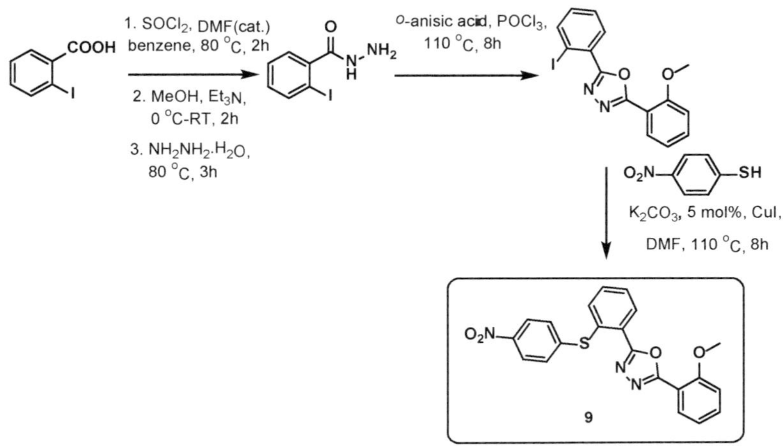

Scheme 10. Synthesis of oxadiazole derivative 9.

3.2 Pd Catalyzed Oxadiazole Synthesis

In order to study the effect of oligomers on semiconducting properties, Lee et al. synthesized a series of oxadiazole scaffolds utilizing coupling chemistry (Lee et al. 2009). These hybrid compounds could be useful for applications of OFET (organic field effect transistor).

The reaction of 4-bromo-4'-iodobiphenyl, in the presence of diphenylamine, induced diphenyl amine capped intermediate in 71% yield. After that, it underwent a coupling reaction with $Pd(PPh_3)_2Cl_2$ and copper(I) iodide (Scheme 11), followed by a reaction with NaOH to produce 4-(N, N-Diphenylamino)phenyl-4'-ethynylbiphenyl in 70% yield (Pålsson et al. 2010). Further coupling with $Pd(PPh_3)_2Cl_2$ in the presence of copper(I) iodide afforded target oxadiazole compound 10 in 42% yield.

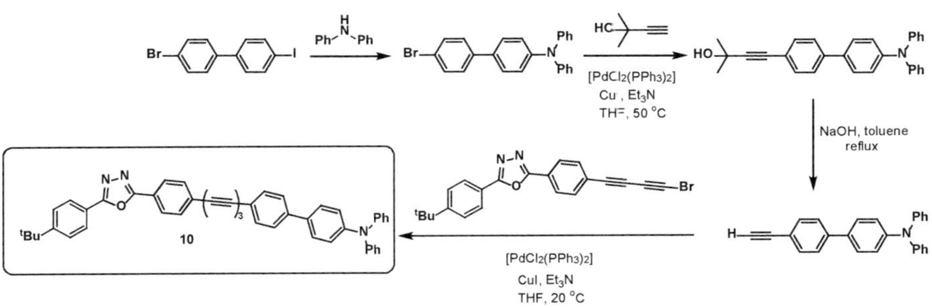

Scheme 11. Synthesis of oxadiazole derivative 10.

3.3 Microwave-Assisted Synthesis

Wang et al. synthesized oxadiazole 11 from a reaction of biphenyl-4-carboxylic acid with benzamide under microwave heating (Scheme 12) at 150°C in 85% yield (Wang et al. 2006). Microwave heating, along with solid-phase reagents, improved the purification process. In 2020, the reaction of isothiocyanates and hydrazides in

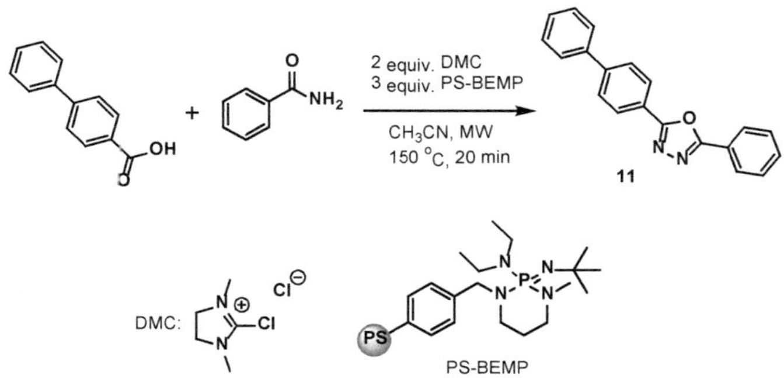

Scheme 12. Synthesis of oxadiazole derivative 11.

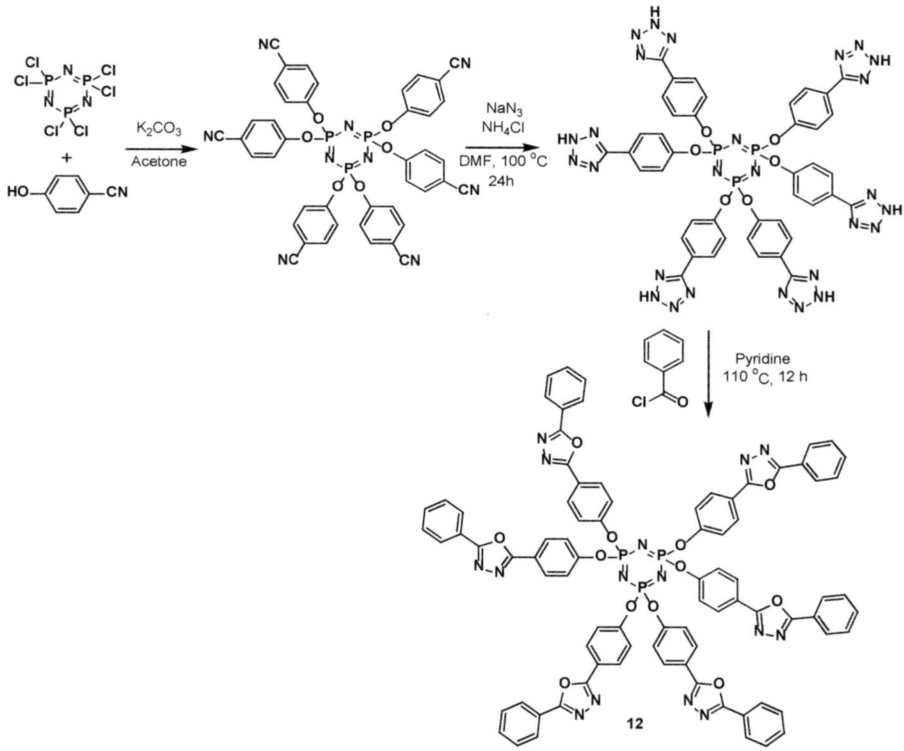

Scheme 13. Synthesis of oxadiazole derivative 12.

the presence of tert-butyl hydroperoxide (TBHP) and microwave heating resulted in 2-amino 1,3,4-oxadiazole in water (Kumar Sigalapalli et al. 2020).

In late 2010, Liu et al. synthesized cyclotriphosphazene-centered oxadiazole 12 in 61% yield through a three-step process (Liu et al. 2011). Hexachlorophosphazene, in the presence of 4-cyanophenol and K_2CO_3, produced an intermediate hexakis(4-cyanophenyl)-cyclotriphosphazene (Scheme 13). The cyano intermediate further reacted in the presence of sodium azide and ammonium chloride to produce hexakis(4-tetrazolylphenyloxy)-cyclotriphosphazene in 89% yield. Finally, the tetrazole intermediate reacted with benzoyl chloride in pyridine to yield the target compound 12.

3.4 Solid Phase-Based Synthesis of 1,3,4-Oxadiazole Derivative

The solid-phase synthetic technique is a rapid preparation tool for a diverse range of chemical products. It usually provides straightforward separation and purification where impurities can be separated through filtration. In 2005, Severinsen et al. developed a method for the construction of alkyl/arylamino-[1,3,4]-oxadiazoles in high yields utilizing resin-bound semicarbazides in three synthetic steps (Severinsen et al. 2005). In 2007, vinyl-substituted oxadiazole derivatives were synthesized using polymer-based selenium resin with high-purity isolated products (Wang et al. 2007). In 2005, Baxendale et al. successfully tailored a series of 5-substituted-2-amino-1,3,4-oxadiazoles from a combination of three different starting materials such as acylhydrazine, sulfonyl chloride, and an isocyanate and in the presence of polymer assisted phosphazine base (Baxendale et al. 2005).

Similar to conventional synthetic methods, resin-bound organic scaffolds can undergo chemical transformation to give a diverse range of useful products. In 2006, Liu et al. synthesized 1,3,4-oxadiazole derivative 13 utilizing resin-bound intermediates (Scheme 14) (Liu et al. 2006). Later coupling of Fmoc-attached amino acid with thiosemicarbazide resin resulted in 1,3,4-oxadiazoles in 2018 (Abdildinova et al. 2018). Derivatization of amino acids on both sides yielded a range of compounds useful for peptidomimetics.

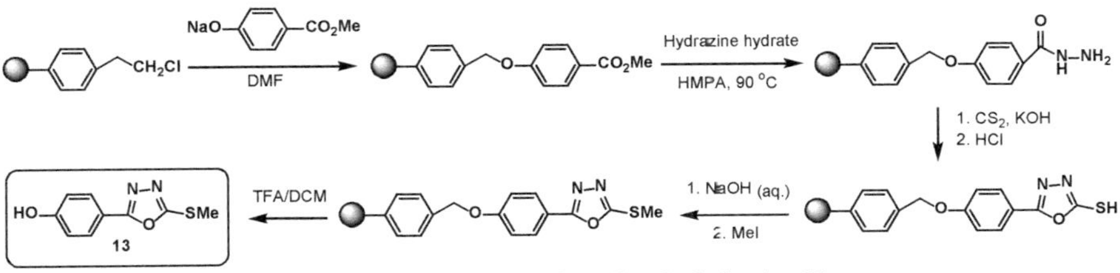

Scheme 14. Synthesis of oxadiazole derivative 13.

3.5 Iodine-Promoted Synthesis of 1,3,4-Oxadiazole Derivative

Cheaper, accessible reagents play a vital role in executing large-scale organic reactions. Molecular iodine is one such reagent that is available at a low cost and plays a significant role in oxadiazole synthesis through C-O bond formation, such as during the reaction of aldehydes with hydrazides.

One-pot cyclization of 4-phenylsemicarbazide with substituted aryl methyl ketone was achieved by Zhang et al. (2022). I_2/DMSO facilitated the cyclization process in an oxidative environment and produced compound 14 in 80% yield with K_2CO_3 as a base. Initially, reaction with I_2 ensured the formation of -iodo ketone (Schemes 15–16).

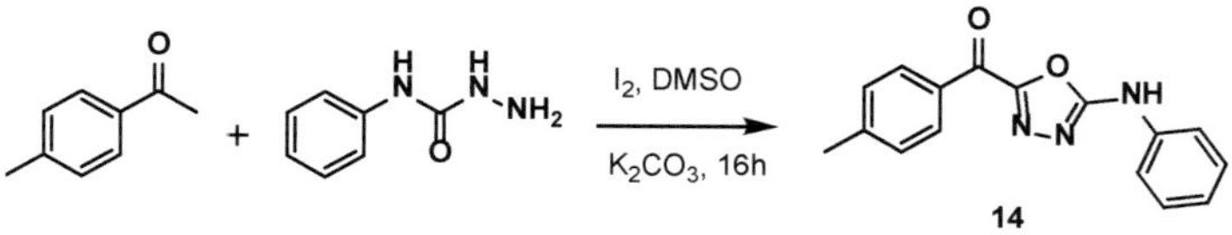

14

Scheme 15. Synthesis of oxadiazole derivative 14.

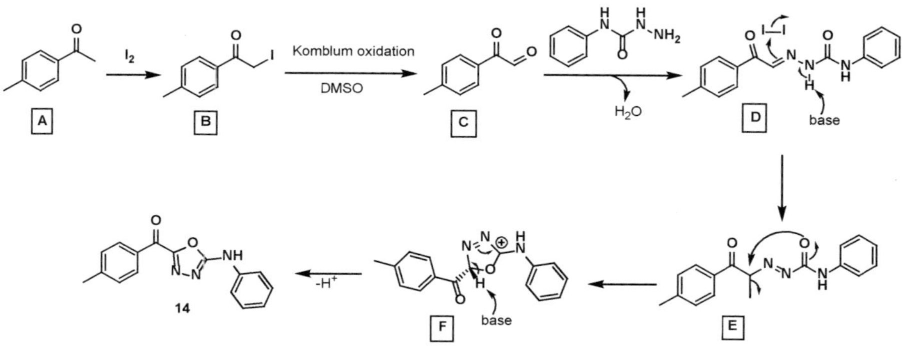

Scheme 16. Mechanism for the synthesis of oxadiazole derivative 14 (Zhang et al. 2022).

3.6 1,3,4-Oxadiazole Synthesis Using Oxidants

Like 1,2,4-oxadiazole synthesis, various oxidizing agents can also facilitate the construction of 1,3,4-oxadiazole derivatives. In 2015, Wang et al. reported the synthesis of 1,3,4-oxadiazole during a reaction of aryl tetrazole and an aryl aldehyde using Di-tert-butyl peroxide (DTBP) as an oxidant (Wang et al. 2015). This one-pot reaction proceeded with N-acylation of tetrazole substituent, and it was subjected to thermal rearrangement to construct the cyclic oxadiazole ring in moderate to high yields. A combination of arylacetic acid with substituted hydrazides in the presence of K_2CO_3 can yield 2,5-disubstituted 1,3,4-oxadiazole scaffolds in an oxygen environment (Lekkala et al. 2022). At first, this reaction undergoes an oxidative decarboxylation process of arylacetic acid and later an oxidative functionalization of imine C-H bonds to generate oxadiazole derivatives in up to 92% yield. *In situ* formation of tert-butyl hypoiodite (*t*-BuOI) from *t*-BuOCl and NaI could be useful for cyclization of N-acylhydrazones to generate substituted oxadiazoles (Gao and Wei 2013).

3.7 Visible Light Promoted 1,3,4-Oxadiazole Synthesis

Visible light is often considered a clean and effective source of energy; the introduction of visible light unravels new chemical reactions with excellent selectivity. In an effort to achieve an efficient method for 1,3,4-oxadiazole synthesis, Kapoorr et al. utilized visible light to transform semicarbazones into oxadiazole 15 in the presence of a popular photoredox catalyst called "eosin Y" (Kapoorr et al. 2015). CBr_4 was used as a brominating source (Schemes 17–18).

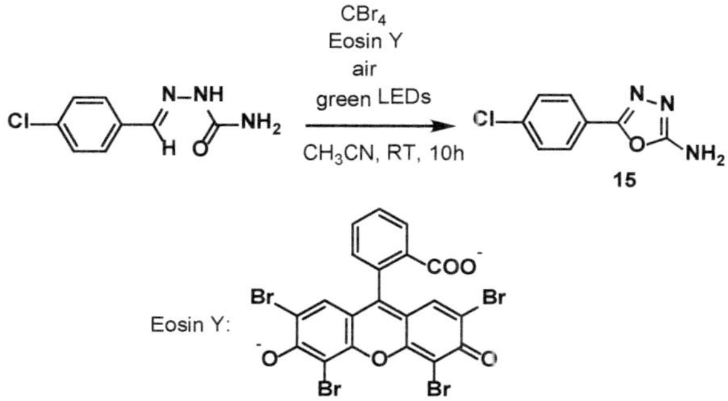

Scheme 17. Synthesis of oxadiazole derivative 15.

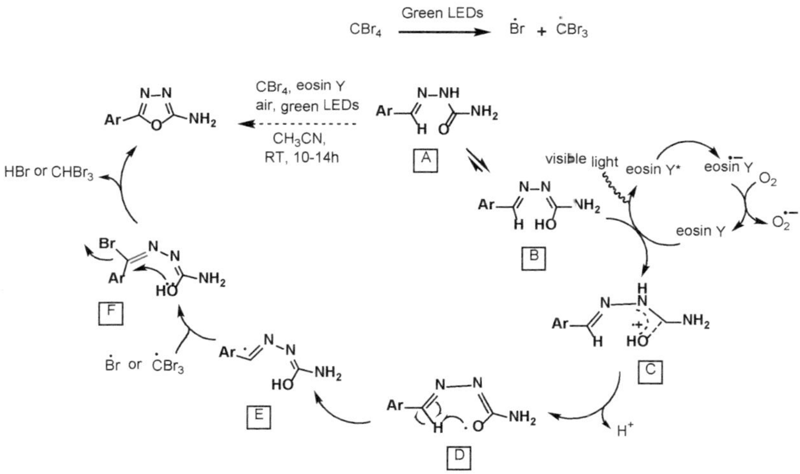

Scheme 18. Mechanism for synthesis of oxadiazole derivative 15 (Kapoorr et al. 2015).

Recently, a combination of aryl aldehyde with hypervalent iodine (III) reagent in the presence of visible light irradiation induced the formation of 2,5-disubstituted 1,3,4-oxadiazole derivatives in up to 89% yields (Li et al. 2021). Interestingly, the reaction requires neither an additional catalyst nor the pre-activation of aldehydes. Applying photocatalyst eosin Y into a solution of hydrazides and substituted diketones resulted in 1,3,4-oxadiazole scaffolds in moderate to good yields (Diao et al. 2018). The introduction of photocatalysts facilitated decarboxylation and cyclization, leading to oxadiazole formation. Visible light can also be useful for transforming multicomponent starting materials into their respective target molecules. In 2021, Russo et al. summarized the preparation of 1,3,4-oxadiazole derivatives utilizing a combination of carboxylic acids, N-alkyl-N-methylanilines, and N-isocyanoiminotriphenylphosphorane in the presence of 30 W blue LED light irradiation. N, N-dimethylaniline was oxidized to iminium cation promoted by a ruthenium photocatalyst. The reaction resulted in the formation of three new bonds (such as C-C, C-O, and C-N bonds) in the single reaction, and it could also tolerate a variety of functional groups (Russo et al. 2021).

3.8 Polymer-Based 1,3,4-Oxadiazole Derivative Synthesis

Thermally stable polymers with no decomposition below 386°C based on oxadiazole motifs were fabricated by Chen et al. (1999). The introduction of poly(p-phenylene vinylene) unit in the polymeric backbone assists in maintaining morphological stability during the preparation of thin films. Later, in 2002, Ding et al. synthesized various copolymers by installing alternating 9,9-dioctylfluorene and oxadiazole units through a Suzuki cross-coupling reaction (Ding et al. 2002). Copolymer samples exhibited blue emission in dichloromethane with a 70% quantum yield.

Kim et al. developed a 9,9'-dioctylfluorene anchored aromatic polyoxadiazole scaffold 20 through a Suzuki coupling reaction (Kim et al. 2006). The authors prepared the monomer unit (oxadiazole) (51%) while mixing 5-bromo-2-hydroxybenzoic acid in the presence of polyphosphoric acid (PPA). The oxadiazole moiety in the polymer was guarded by the bis(hydroxyphenyl) group in its 2-position (Scheme 19). The tert-butoxycarbonyl group was utilized to protect the highly vulnerable hydroxyl group before polymerization and subsequently eliminated

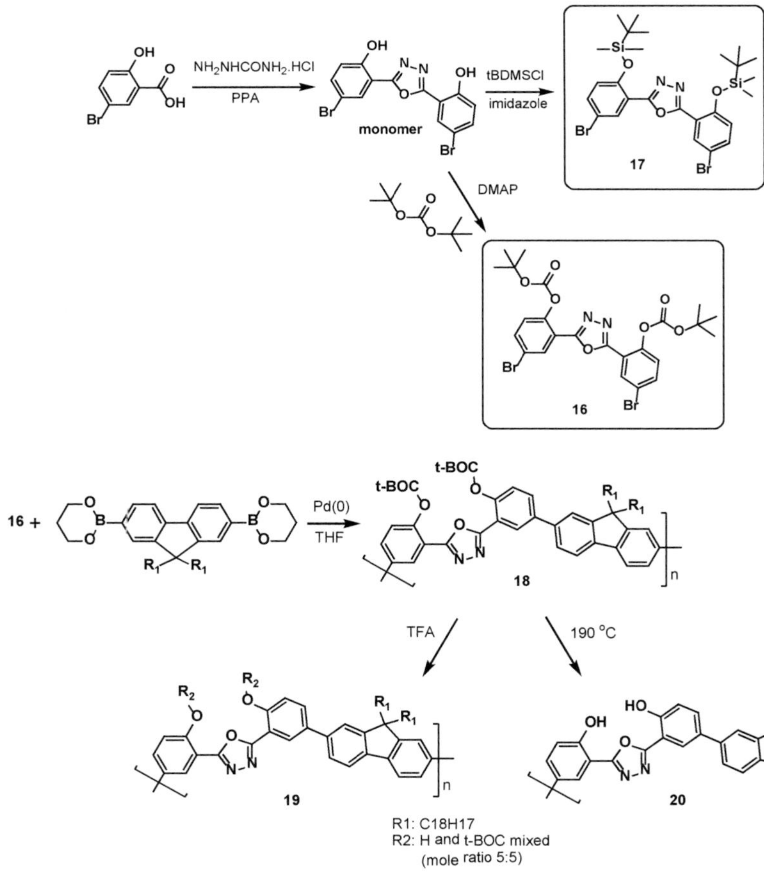

Scheme 19. Synthesis of oxadiazole derivatives 16–20.

post-polymerization with trifluoroacetic acid or at high temperature (190°C). The existence of bis(hydroxyphenyl) moiety imparted high Stokes shift owing to ESIPT (excited state intramolecular proton transfer).

3.9 Other Methods for Synthesis of 1,3,4-Oxadiazole Derivative

β-Glucuronidase is an enzyme that catalyzes the transformation of β-glucuronoside to D-glucuronate and an alcohol. Taha et al. synthesized oxadiazole derivative 21 in four synthetic steps with a 94% yield (Taha et al. 2015). Methyl salicylate formed an intermediate hydrazone while reacting in the presence of hydrazine in methanol. The hydrazone intermediate reacted with methyl 4-formylbenzoate and a catalytic amount of CH_3COOH in methanol, and subsequent reaction with $PhI(OAc)_2$ produced oxadiazole intermediate (Scheme 20). (Diacetoxyiodo)benzene serves as an oxidant and facilitates intramolecular cyclization to yield oxadiazole. The presence of the hydroxyl group on the benzene group played a major role in inhibiting β-glucuronidase.

Similarly, in another study, (diacetoxyiodo)benzene was utilized for intramolecular cyclization by Taha et al. and synthesized oxadiazole derivative 22 (Scheme 21) (Taha et al. 2015).

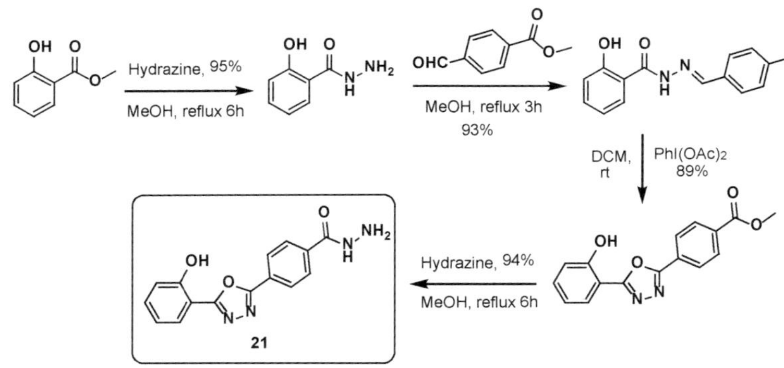

Scheme 20. Synthesis of oxadiazole derivative 21.

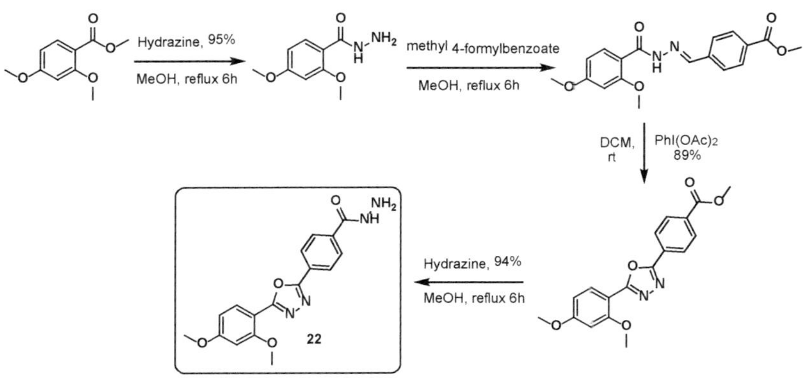

Scheme 21. Synthesis of oxadiazole derivative 22.

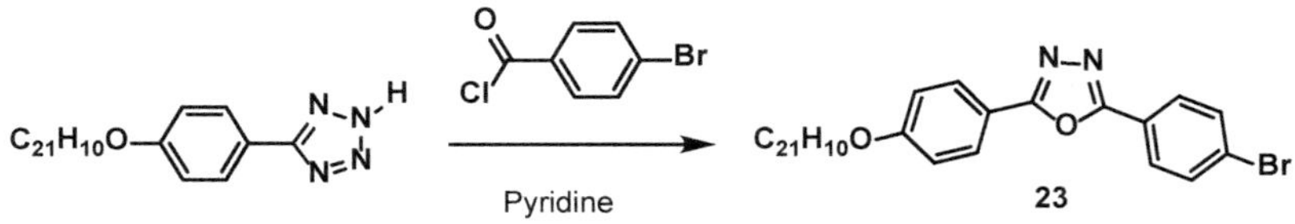

Scheme 22. Synthesis of oxadiazole derivative 23.

To develop luminescent liquid crystals, Cristiano et al. synthesized unsymmetrical oxadiazole derivative 23 utilizing substituted tetrazole (Scheme 22) and the corresponding 4-bromo benzoyl chloride in pyridine solvent (Cristiano et al. 2006).

4. Conclusions

Construction of a 5-membered heterocyclic scaffold such as 1,2,4-oxadiazole in a cost-effective process holds promising potential as it is heavily utilized as a stable amide and ester bioisosteres in peptide mimetics. The chemical and biological behavior of 1,2,4-oxadiazole scaffolds have good potential in drug development. Most notable examples include commercially available drugs with a 1,2,4-oxadiazole motif, such as oxolamine (cough suppressant), Prenoxdiazine, Fasiplon (medication for anxiety reduction), Pleconaril (antiviral drug), Ataluren, Proxazole (analgesic and anti-inflammatory drug), etc. Several new compounds with 1,2,4-oxadiazole motifs are in clinical trials.

Substituted 1,3,4-oxadiazoles are considered functional surrogates of esters, carboxamides, and acids. Also, 1,3,4-oxadiazole compounds are present in pharmaceuticals as anti-inflammatory, anticancer, antiallergic, antimalarial, antiviral, and several natural product-based compounds. These compounds can be prepared by cyclizing N-acyl hydrazones using various oxidants or dehydrative cyclization using $POCl_3$, thionyl chloride, sulfuric acid, etc. Additionally, introducing visible light, microwave technology, etc., for organic synthesis is a powerful strategy owing to faster and more straightforward reactions.

References

Abdildinova, A., S.-J. Yang and Y.-D. Gong. 2018. Solid-phase parallel synthesis of 1, 3, 4-oxadiazole-based peptidomimetic library as a potential modulator of protein-protein interactions. Tetrahedron 74(6): 684–691.

Andersen, T. L., W. Caneschi, A. Ayoub, A. T. Lindhardt, M. R. Couri and T. Skrydstrup. 2014. 1, 2, 4- and 1, 3, 4-Oxadiazole synthesis by palladium-catalyzed carbonylative assembly of aryl bromides with amidoximes or hydrazides. Adv. Synth. Catal. 356(14-15): 3074–3082.

Augustine, J. K., V. Vairaperumal, S. Narasimhan, P. Alagarsamy and A. Radhakrishnan. 2009. Propylphosphonic anhydride (T3P®): an efficient reagent for the one-pot synthesis of 1, 2, 4-oxadiazoles, 1, 3, 4-oxadiazoles, and 1, 3, 4-thiadiazoles. Tetrahedron 65(48): 9989–9996.

Baxendale, I. R., S. V. Ley and M. Martinelli. 2005. The rapid preparation of 2-aminosulfonamide-1, 3, 4-oxadiazoles using polymer-supported reagents and microwave heating. Tetrahedron 61(22): 5323–5349.

Boström, J., A. Hogner, A. Llinàs, E. Wellner and A. T. Plowright. 2012. Oxadiazoles in medicinal chemistry. J. Med. Chem. 55(5): 1817–1830.

Cai, B.-G., Z.-L. Chen, G.-Y. Xu, J. Xuan and W.-J. Xiao. 2019. [3+ 2]-Cycloaddition of 2 H-azirines with nitrosoarenes: visible-light-promoted synthesis of 2, 5-dihydro-1, 2, 4-oxadiazoles. Org. Lett. 21(11): 4234–4238.

Carroll, F. I., J. L. Gray, P. Abraham, M. A. Kuzemko, A. H. Lewin, J. W. Boja et al. 1993. 3-Aryl-2-(3'-substituted-1', 2', 4'-oxadiazol-5'-yl) tropane analogs of cocaine: affinities at the cocaine binding site at the dopamine, serotonin, and norepinephrine transporters. J. Med. Chem. 36(20): 2886–2890.

Chen, Z.-K., H. Meng, Y.-H. Lai and W. Huang. 1999. Photoluminescent poly (p-phenylenevinylene) s with an aromatic oxadiazole moiety as the side chain: synthesis, electrochemistry, and spectroscopy study. Macromolecules 32(13): 4351–4358.

Cristiano, R., A. A. Vieira, F. Ely and H. Gallardo. 2006. Synthesis and characterization of luminescent hockey stick-shaped liquid crystalline compounds. Liq. Cryst. 33(04): 381–390.

Deegan, T. L., T. J. Nitz, D. Cebzanov, D. E. Pufko and J. A. Forco Jr. 1999. Parallel synthesis of 1, 2, 4-oxadiazoles using CDI activation. Bioorg. Med. Chem. Lett. 9(2): 209–212.

Diao, P., Y. Ge, C. Xu, N. Zhang and C. Guo. 2018. Synthesis of 2, 5-disubstituted 1, 3, 4-oxadiazoles by visible-light-mediated decarboxylation–cyclization of hydrazides and diketones. Tetrahedron Lett. 59(8): 767–770.

Ding, J., M. Day, G. Robertson and J. Roovers. 2002. Synthesis and characterization of alternating copolymers of fluorene and oxadiazole. Macromolecules 35(9): 3474–3483.

Gao, P. and Y. Wei. 2013. Efficient oxidative cyclization of N-acylhydrazones for the synthesis of 2, 5-disubstituted 1, 3, 4-oxadiazoles using t-BuOI under neutral conditions. Heterocycl. Commun. 19(2): 113–119.

Gómez-Saiz, P., J. García-Tojal, M. A. Maestro, F. J. Arnaiz and T. Rojo. 2002. Evidence of desulfurization in the oxidative cyclization of thiosemicarbazones. Conversion to 1, 3, 4-oxadiazole derivatives. Inorg. Chem. 41(6): 1345–1347.

Hébert, N., A. L. Hannah and S. C. Sutton. 1999. Synthesis of oxadiazoles on solid support. Tetrahedron Lett. 40(49): 8547–8550.

Hu, Q. S., S. R. Sheng, X. L. Liu, F. Hu and M. Z. Cai. 2008. Facile solid-phase organic synthesis of 5-vinyl-substituted 1, 2, 4-oxadiazoles from polymer-bound α-selenopropionic acid. J. Chin. Chem. Soc. 55(4): 768–771.

Jagadeesh Prathap, K., M. Himaja, S. V. Mali and D. Munirajasekhar. 2014. Synthesis of new (Pyrazol-3-yl)-1, 3, 4-oxadiazole derivatives by unexpected aromatization during oxidative cyclization of 4, 5-dihydro-1h-pyrazole-3-carbohydrazones and their biological activities. J. Heterocycl. Chem. 51(3): 726–732.

Kaplancikli, Z. A. 2011. Synthesis of some oxadiazole derivatives as new anticandidal agents. Molecules 16(9): 7662–7671.

Kapoorr, R., S. N. Singh, S. Tripathi and L. D. S. Yadav. 2015. Photocatalytic oxidative heterocyclization of semicarbazones: an efficient approach for the synthesis of 1, 3, 4-oxadiazoles. Synlett 26(09): 1201–1206.

Kim, T. H., H. J. Kim, C. G. Kwak, W. H. Park and T. S. Lee. 2006. Aromatic oxadiazole-based conjugated polymers with excited-state intramolecular proton transfer: Their synthesis and sensing ability for explosive nitroaromatic compounds. J. Polym. Sci., Part A: Polym. Chem. 44(6): 2059–2068.

Kumar Sigalapalli, D., M. Kadagathur, A. Sujat Shaikh, G. S. Jadhav, B. Bakchi, B. Nagendra Babu et al. 2020. Microwave-assisted TBHP-mediated synthesis of 2-amino-1, 3, 4-oxadiazoles in water. ChemistrySelect 5(42): 13248–13258.

Kuram, M. R., W. G. Kim, K. Myung and S. Y. Hong. 2016. Copper-Catalyzed Direct Synthesis of 1, 2, 4-Oxadiazoles from Amides and Organic Nitriles by Oxidative N–O Bond Formation, Wiley Online Library.

Lee, T., C. A. Landis, B. M. Dhar, B. J. Jung, J. Sun, A. Sarjeant et al. 2009. Synthesis, structural characterization, and unusual field-effect behavior of organic transistor semiconductor oligomers: inferiority of oxadiazole compared with other electron-withdrawing subunits. J. Am. Chem. Soc. 131(5): 1692–1705.

Lekkala, C., V. Bodala, K. Yettula, B. K. Karasala, R. L. Podugu and S. Vidavalur. 2022. Copper-catalyzed one-pot synthesis of 2, 5-disubstituted 1, 3, 4-oxadiazoles from arylacetic acids and hydrazides via dual oxidation. ACS Omega 7(31): 27157–27163.

Li, J., J.-X. Wen, X.-C. Lu, G.-Q. Hou, X. Gao, Y. Li et al. 2021. Catalyst-free visible-light-promoted cyclization of aldehydes: access to 2, 5-disubstituted 1, 3, 4-oxadiazole derivatives. ACS Omega 6(40): 26699–26706.

Liang, G.-B. and D. D. Feng. 1996. An improved oxadiazole synthesis using peptide coupling reagents. Tetrahedron Lett. 37(37): 6627–6630.

Liu, S.-Z., X. Wu, A.-Q. Zhang, J.-J. Qiu and C.-M. Liu. 2011. 1D nano- and microbelts self-assembled from the organic-inorganic hybrid molecules: oxadiazole-containing cyclotriphosphazene. Langmuir 27(7): 3982–3990.

Liu, Z., J. Zhao and X. Huang. 2006. Solid-phase synthesis of 1, 3, 4-oxadiazoline-5-thione derivatives from resin-bound acylhydrazines. Bioorg. Med. Chem. Lett. 16(7): 1828–1830.

Makara, G. M., P. Schell, K. Hanson and D. Moccia. 2002. An efficient solid-phase synthesis of 3-alkylamino-1, 2, 4-oxadiazoles. Tetrahedron Lett. 43(29): 5043–5045.

Mansoori, Y., G. Barghian, B. Koohi-Zargar, G. Imanzadeh and M. Zamanloo. 2012. Thermally stable polymers containing 1, 3, 4-oxadiazole units obtained from Huisgen reaction. Chin. J. Polym. Sci. 30: 36–44.

Mohammed, I., I. R. Kummetha, G. Singh, N. Sharova, G. Lichinchi, J. Dang et al. 2016. 1, 2, 3-Triazoles as amide bioisosteres: discovery of a new class of potent HIV-1 Vif antagonists. J. Med. Chem. 59(16): 7677–7682.

Nandeesh, K. N., H. A. Swarup, N. C. Sandhya, C. D. Mohan, C. S. P. Kumar, M. N. Kumara et al. 2016. Synthesis and antiproliferative efficiency of novel bis (imidazol-1-yl) vinyl-1, 2, 4-oxadiazoles. New J. Chem. 40(3): 2823–2828.

Niu, P., J. Kang, X. Tian, L. Song, H. Liu, J. Wu et al. 2015. Synthesis of 2-amino-1, 3, 4-oxadiazoles and 2-amino-1, 3, 4-thiadiazoles via sequential condensation and I2-mediated oxidative C–O/C–S bond formation. J. Org. Chem. 80(2): 1018–1024.

Orlek, B. S., F. E. Blaney, F. Brown, M. S. Clark, M. S. Hadley, J. Hatcher et al. 1991. Comparison of azabicyclic esters and oxadiazoles as ligands for the muscarinic receptor. J. Med. Chem. 34(9): 2726–2735.

Pace, A. and P. Pierro. 2009. The new era of 1, 2, 4-oxadiazoles. Org. Biomol. Chem. 7(21): 4337–4348.

Pålsson, L. O., C. Wang, A. S. Batsanov, S. M. King, A. Beeby, A. P. Monkman et al. 2010. Efficient intramolecular charge transfer in oligoyne-linked donor–π–acceptor molecules. Chem. Eur. J. 16(5): 1470–1479.

Parker, P. D. and J. G. Pierce. 2016. Synthesis of 1, 2, 4-oxadiazoles via DDQ-mediated oxidative cyclization of amidoximes. Synthesis 48(12): 1902–1909.

Russo, C., R. Cannalire, P. Luciano, F. Brunelli, G. C. Tron and M. Giustiniano. 2021. Visible-light photocatalytic ugi/aza-wittig cascade towards 2-aminomethyl-1, 3, 4-oxadiazole derivatives. Synthesis 53(23): 4419–4427.

Sams, C. K. and J. Lau. 1999. Solid-phase synthesis of 1, 2, 4-oxadiazoles. Tetrahedron Lett. 40(52): 9359–9362.

Severinsen, R., J. P. Kilburn and J. F. Lau. 2005. Versatile strategies for the solid phase synthesis of small heterocyclic scaffolds:[1, 3, 4]-thiadiazoles and [1, 3, 4]-oxadiazoles. Tetrahedron 61(23): 5565–5575.

Shaikh, A. L. N., M. Bhoye, Y. Nandurkar, H. R. Pawar, V. Bobade and P. C. Mhaske. 2023. An efficient synthesis of new 3, 5-bis (2-arylthiazol-4-yl)-1, 2, 4-oxadiazole derivatives and their antimicrobial evaluation. J. Heterocycl. Chem.

Swain, C., R. Baker, C. Kneen, J. Moseley, J. Saunders, E. Seward et al. 1991. Novel 5-HT3 antagonists. Indole oxadiazoles. J. Med. Chem. 34(1): 140–151.

Taha, M., N. H. Ismail, S. Imran, M. Q. B. Rokei, S. M. Saad and K. M. Khan. 2015. Synthesis of new oxadiazole derivatives as α-glucosidase inhibitors. Bioorg. Med. Chem. 23(15): 4155–4162.

Taha, M., N. H. Ismail, S. Imran, M. Selvaraj, A. Rahim, M. Ali et al. 2015. Synthesis of novel benzohydrazone–oxadiazole hybrids as β-glucuronidase inhibitors and molecular modeling studies. Bioorg. Med. Chem. 23(23): 7394–7404.

Tiemann, F. and P. Krüger. 1884. Ueber amidoxime und azoxime. Ber. Dtsch. Chem. Ges. 17(2): 1685–1698.

Tokumaru, K. and J. N. Johnston. 2017. A convergent synthesis of 1, 3, 4-oxadiazoles from acyl hydrazides under semiaqueous conditions. Chem. Sci. 8(4): 3187–3191.

Wang, C., G.-Y. Jung, Y. Hua, C. Pearson, M. R. Bryce, M. C. Petty et al. 2001. An efficient pyridine-and oxadiazole-containing hole-blocking material for organic light-emitting diodes: synthesis, crystal structure, and device performance. Chem. Mater. 13(4). 1167–1173.

Wang, L., J. Cao, Q. Chen and M. He. 2015. One-pot synthesis of 2, 5-diaryl 1, 3, 4-oxadiazoles via di-tert-butyl peroxide promoted N-acylation of aryl tetrazoles with aldehydes. J. Org. Chem. 80(9): 4743–4748.

Wang, Q., K. C. Mgimpatsang, M. Konstantinidou, S. V. Shishkina and A. Dömling. 2019. 1, 3, 4-Oxadiazoles by Ugi-tetrazole and Huisgen reaction. Org. Lett. 21(18): 7320–7323.

Wang, Y.-G., W.-M. Xu and X. Huang. 2007. Selenium-based safety-catch linker: solid-phase synthesis of vinyl-substituted oxadiazoles and triazoles. J. Comb. Chem. 9(3): 513–519.

Wang, Y., D. R. Sauer and S. W. Djuric. 2006. A simple and efficient one step synthesis of 1, 3, 4-oxadiazoles utilizing polymer-supported reagents and microwave heating. Tetrahedron Lett. 47(1): 105–108.

Watjen, F., R. Baker, M. Engelstoff, R. Herbert, A. MacLeod, A. Knight et al. 2002. Novel benzodiazepine receptor partial agonists: oxadiazolylimidazobenzodiazepines. J. Med. Chem. 32(10): 2282–2291.

Yang, S.-J., J.-M. Lee, G.-H. Lee, N. Kim, Y.-S. Kim and Y.-D. Gong. 2014. Microwave assisted synthesis of 1, 3, 4-oxadiazole/thiohydantoin hybrid derivatives via dehydrative cycliztion of semicarbazide. Bull. Korean Chem. Soc. 35: 3609–3617.

Zhang, F.-L., Y.-F. Wang and S. Chiba. 2013. Orthogonal aerobic conversion of N-benzyl amidoximes to 1, 2, 4-oxadiazoles or quinazolinones. Org. Biomol. Chem. 11(36): 6003–6007.

Zhang, S., C. Liu, X. Wu, W. Li, H. Li, S. Wang et al. 2022. Synthesis of 1, 3, 4-oxadiazoles by iodine-mediated oxidative cyclization of methyl ketones with 4-phenylsemicarbazide. Synlett 33(03): 269–272.

CHAPTER 2

1,2,4 and 1,3,4-Oxadiazole Scaffolds in Designing Organic Light-Emitting Diodes

1. Introduction

Solid-state lighting materials, full-color flat panel displays, and back-lit liquid crystal displays are some trending targets where organic light-emitting diodes are used as raw materials. Secondly, organic light-emitting diodes or OLEDs are abundant in mobile phones. Therefore, much research is going on to develop highly efficient, flexible, thin displays with light weight at lower cost. Several parameters, such as stability, energy levels, fluorescence behavior, 100% red/green/blue emission colors, etc., are being tuned accordingly.

Apart from many heterocyclic scaffolds, oxadiazole derivatives are being used tremendously to improve the performance of OLED devices. Materials made with oxadiazole moiety exhibit electron-deficient characteristics, high luminescence quantum yields, and adequate thermal stability. These kinds of available characteristics strengthen oxadiazole-based materials toward better OLED devices with superior electron mobility. Ideal OLED devices possess (a) a steady amount of charge introduction from the two electrodes and (b) systematic transport of holes as well as electrons inside luminescent layers of the device (Hughes and Bryce 2005).

Introducing electron-deficient small molecule dopants in the device can augment the electron transport process. Similarly, the attachment of electron transporting, hole blocking layer alongside the cathode can boost the electron transport performance of the device. Tang and VanSlyke first published a report on OLEDs and their application in flat panel displays in 1987 (Tang and VanSlyke 1987). Later, many heterocyclic compounds were incorporated for tethering OLED materials with suitable optical and electronic properties. For example, 2,5-Diaryl-1,3,4-oxadiazoles with various small molecules and polymers as side chains often behave as good electron transporters (Paun et al. 2016). Alternatively, conjugated polymers such as poly(fluorene) and poly(phenylenevinylene) are well-known hole-transporters (Hughes and Bryce 2005). This chapter summarizes OLED materials originating from two different classes of oxadiazole systems such as 1,2,4-oxadiazole and 1,3,4-oxadiazole.

2. Organic Light-Emitting Diodes Based on 1,2,4-Oxadiazole Derivatives

Oxadiazole motifs exhibit high electron mobility and excellent thermal stability. Materials with high triplet energy are more advantageous for energy transfer to blue triplet emitters, hence useful while fabricating phosphorescent OLEDs (Lee et al. 2014). Li et al. developed two bipolar host materials, 1 and 2, through four synthetic steps utilizing carbazole precursor and 1,2,4-oxadiazole as an n-type (electron withdrawing) group (Li et al. 2014). Both the materials displayed glass transition of more than 100°C in differential scanning calorimetry (DSC), which shows better stability. Oxadiazole materials 1 and 2 exhibited high triplet energy at 2.71 and 2.81 eV, respectively. Development of high current efficiency of 13.0 and 16.0 cd A^{-1} for the synthesized materials 1 and 2 respectively indicates good performance in terms of blue phosphorescent OLEDs.

Taking advantage of the high triplet energies of the 1,2,4-oxadiazole motif, the same group at Soochow University studied organic light-emitting diodes utilizing materials 3 and 4 (Li et al. 2014). The asymmetrically positioned oxadiazole material 3 exhibited good performance with a current efficiency of 22.5 cd A^{-1}, whereas material 4 displayed good performance with a current efficiency of around 19.9 cd A^{-1}.

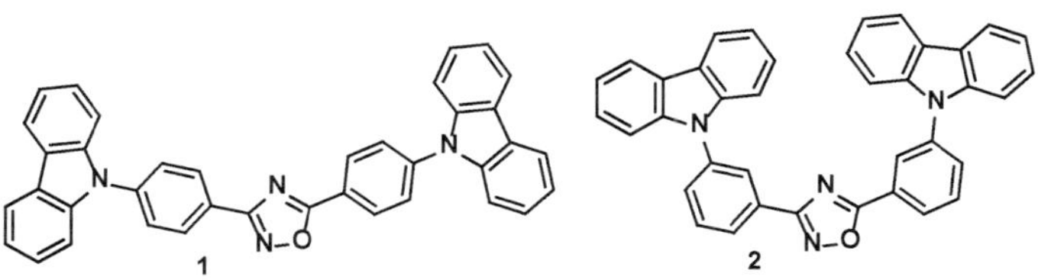

Structures 1 and 2

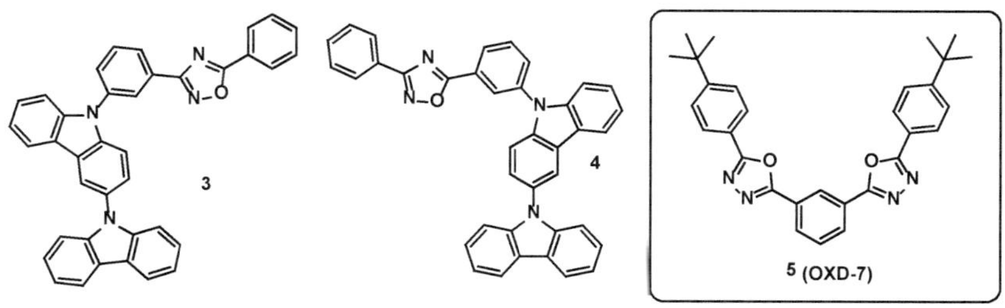

Structures 3, 4 and 5

3. Organic Light-Emitting Diodes Based on 1,3,4-Oxadiazole Derivatives

OXD-7 or 1,3-bis[2-(4-tert-butylphenyl)-1,3,4-oxadiazo-5-yl]benzene is the most common commercially available electron-transporting system based on 1,3,4-oxadiazole units (Song et al. 2017). Tamoto et al. developed several oxadiazole-based dimers and trimer derivatives 6–11 (Tamoto et al. 1997). The attached triphenylamine acted as a hole transport unit, and the corresponding oxadiazole motif served as an electron transport unit.

Wang et al. studied light-emitting properties of bis(1,3,4-oxadiazole) based materials by utilizing poly[2-methoxy-5-(2-ethylhexoxy)-1,4-phenylenevinylene] as emissive material (compound 12–15) (Wang et al. 2001). Fabrication with auxiliary buffer layers might enhance the stability of the material.

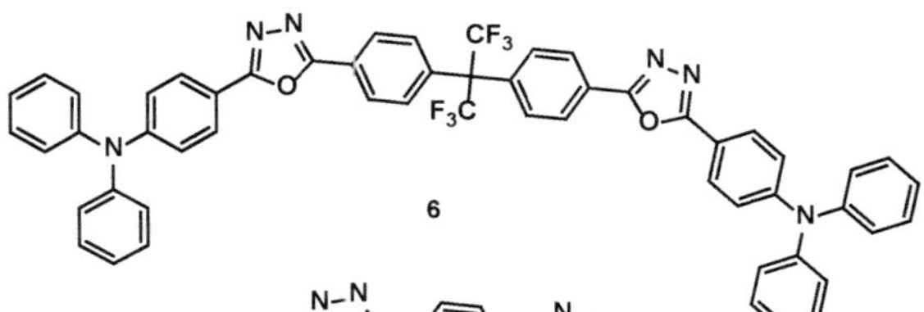

6

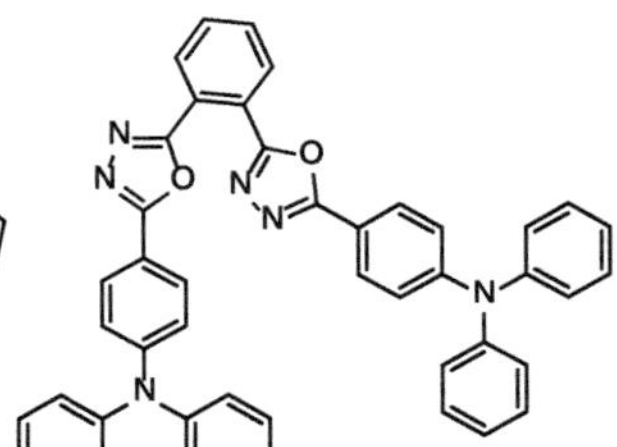

9

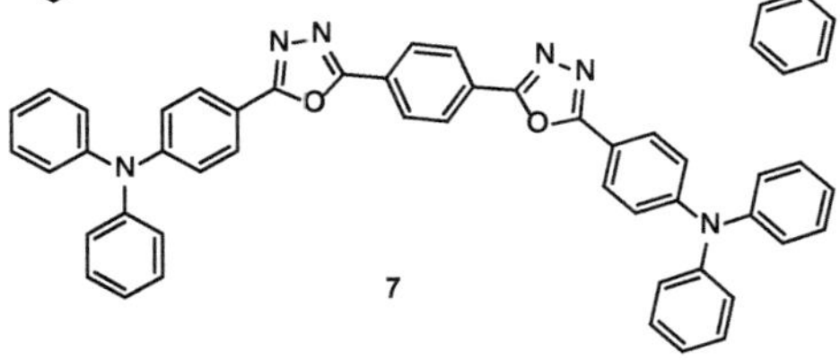

7

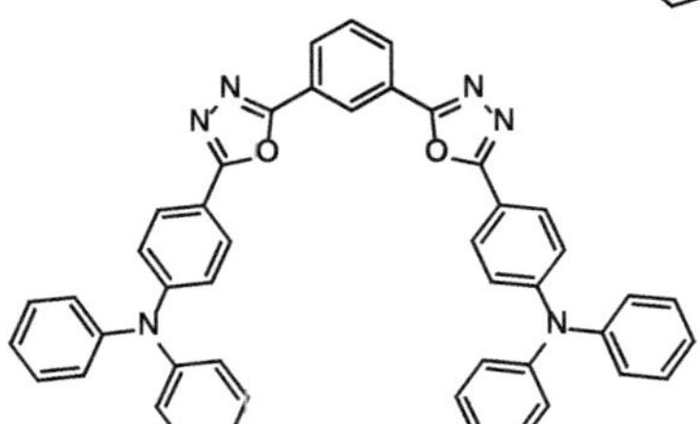

8

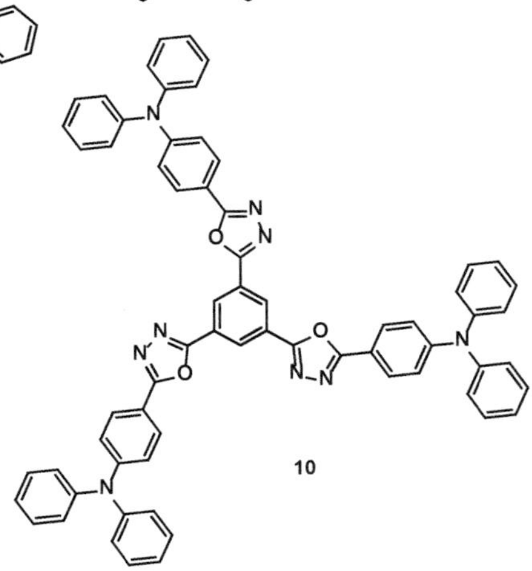

10

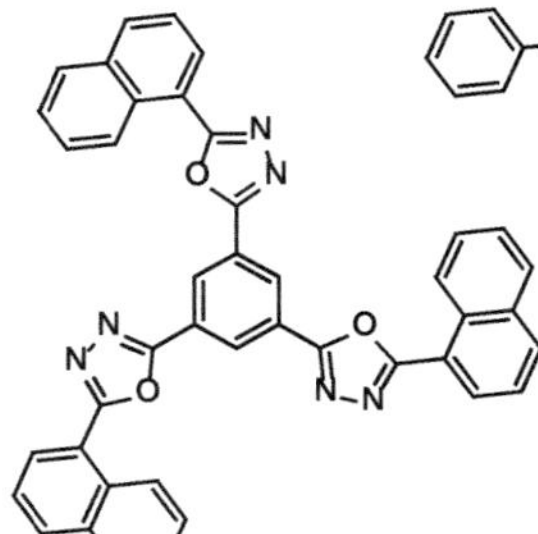

11

Structures 6–11

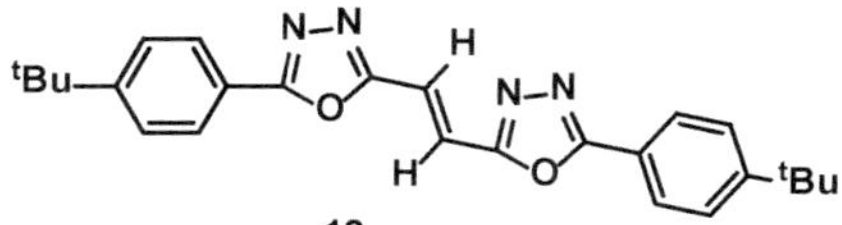

12

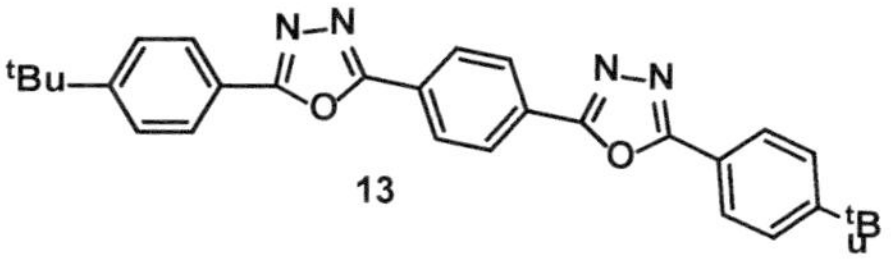

13

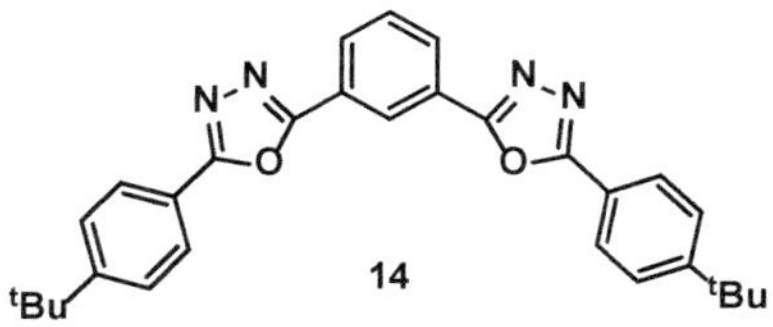

14

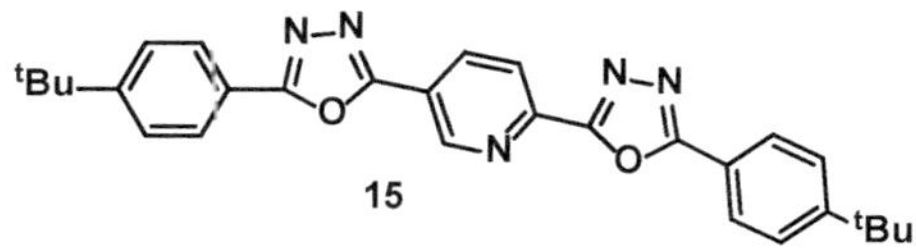

15

Structures 12–15

Chan et al. utilized triphenylsilane-attached oxadiazole 16 to develop high-performance OLEDs (Chan et al. 2002). Five different devices were constructed using compound 16 as a blue light-emitting material. Various parameters such as electroluminescence intensity, efficiency, current density, and blue color purity altered significantly while varying device thickness and electron-transporting material such as tris-(8-hydroxyquinoline) aluminum (Alq_3). One of the devices displayed the highest electroluminescence at 19,000 cd/m^2. The device exhibited large external quantum efficiency (2.4%) and small current density (674 mA/cm^2). Interestingly, the device can withstand temperatures over 100°C without losing luminance.

The efficiency of OLEDs can be enhanced in different ways. One such process is through blended layers. Ahn et al. prepared an OLED device combining an oxadiazole-based electron-transporting material 17 and poly[2-(2-ethylhexyloxy)-5-methoxy-1,4-phenylenevinylene (MEH-PPV) film (Ahn et al. 2004). The addition of 17 as electron transport material improved external quantum efficiencies significantly.

Wu et al. developed a fluorene-based copolymer 18 utilizing electron-deficient oxadiazole and electron-rich triphenylamine moiety (Wu et al. 2005). The copolymer displayed good thermal stability and excellent glass-transition temperature (306°C). It also exhibited bright blue emission in thin film and in a dilute solution, indicating a promising blue light-emitting response. A high brightness of around 7,128 cd/m^2 and a high luminance efficiency of 2.07 cd/A was observed when this copolymer 18 was introduced into the OLED device. Os(fppz) is a highly efficient red-emitting dye consisting of an osmium(II) pyridyl pyrazolate complex (Tung et al. 2004). Incorporation of 2.6 wt % Os(fppz) complex as doped material into the OLED device

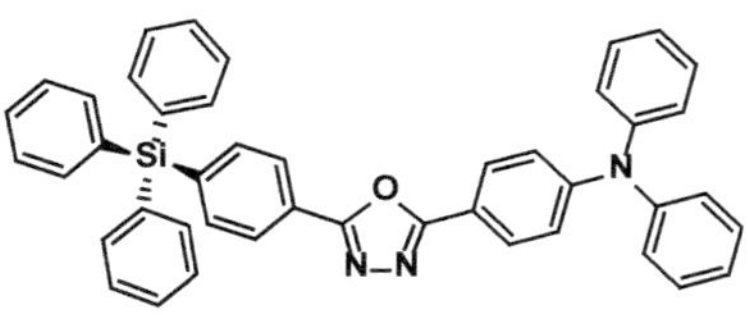

16

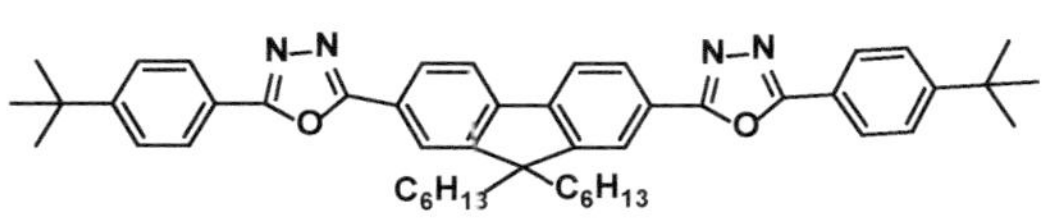

17

Structures 16–17

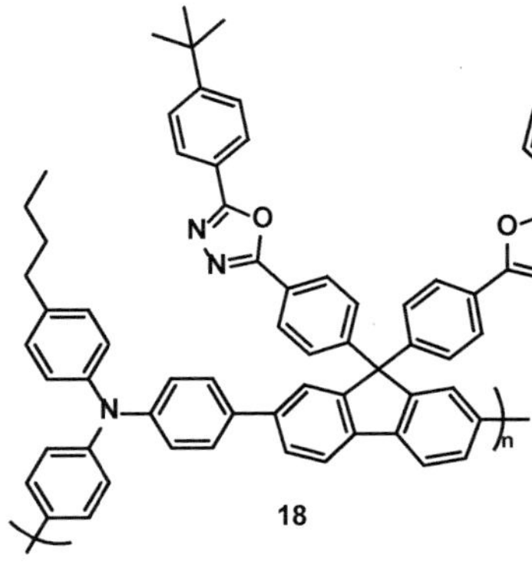

18

Structure 18

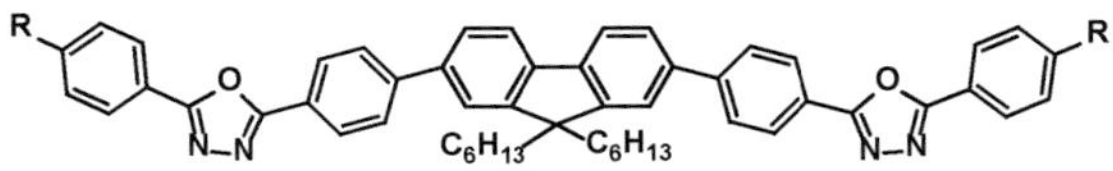

19: R = ^{t}Bu
20: R = $OC_{12}H_{25}$

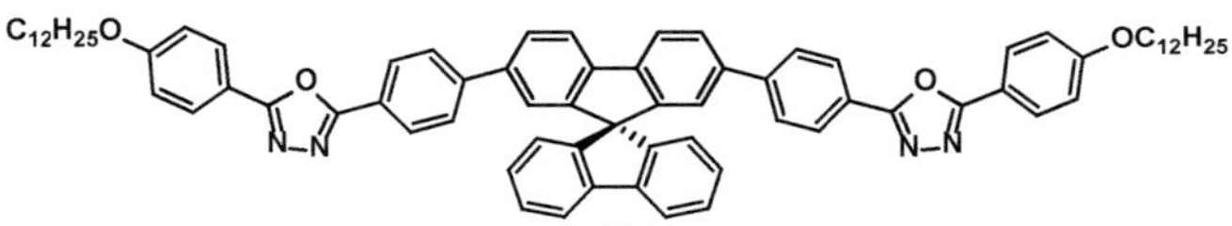

21

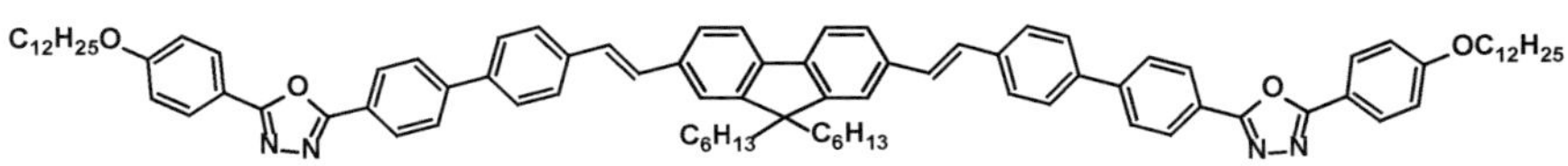

22

Structures 19–22

with copolymer 18 enhanced brightness (28,440 cd/m^2) and external efficiency (9.30%) significantly.

π-conjugated materials display interesting optical and electronic properties and hence contribute significantly to the field of organic electronics (Yuan et al. 2013; Fukushima 2018). Oyston et al. studied π-conjugated materials 19–22 that correspond to linearly designed structures (Oyston et al. 2005). These conjugated materials were synthesized from 9,9-dihexylfluorene or spirobifluorene precursors and aryl/diaryl-oxadiazole motifs. The authors developed OLED devices by blending MEH-PPV as emissive material and the corresponding π-conjugated compounds 19–22 as electron transport materials. The importance of these fabricated devices lies in its fact that MEH-PPV contributes singlehandedly to the total electroluminescence even in the presence of 95% π-conjugated compounds by weight. However, rise in π-conjugated compounds up to 95% by weight enhanced external quantum efficiencies more than two orders of magnitude.

In 2010, Tao et al. utilized Suzuki cross-coupling reactions to synthesize hybrid compounds 23–25 starting from 1,3,4-oxadiazole precursor and 9,9'-spirobifluorene

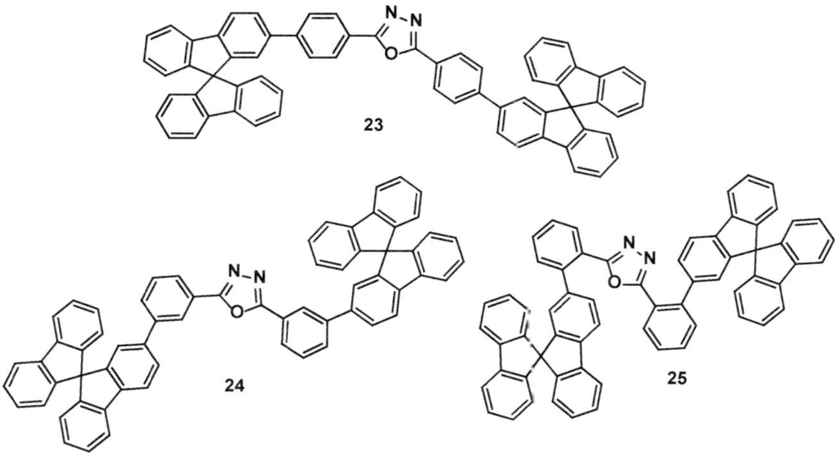

Structures 23–25

derivatives (Tao et al. 2010). Interesting observations were found in terms of thermal as well as morphological stability. Around 136–210°C as glass-transition temperature and 401–480°C as decomposition temperature were noticed for the fabricated devices. The EL performances of the OLED devices constructed from the hybrid compounds followed the sequences 25 > 24 > 23. Out of the three devices, the ortho-linked device (25) displayed the highest external quantum efficiency (9.8%) and highest current efficiency (7.6 cd/A). Structural characteristics, such as twisted configuration, high triplet energy, and separation of HOMO-LUMO energy level favor superior EL performances in the case of hybrid compound 25.

Liou et al. developed aromatic poly(amine-1,3,4-oxadiazole)s 26 and 27 from substituted dicarboxylic acid and terephthalic dihydrazide and (or) isophthalic dihydrazide followed by thermal cyclodehydration (Liou et al. 2006). Both thermal stability and solubility improved significantly due to a bulky naphthalene group in poly(amine-1,3,4-oxadiazole)s. Polymer 26 displayed blue fluorescence with an emission maximum at 463 nm in N-methyl-2-pyrrolidinone solution with a high quantum yield (32%). The polymeric materials also exhibited high glass transition temperatures (263277°C).

Tao et al. obtained hybrid oxadiazole-carbazole product 28 (with 80% yield) during a reaction of 2,5-diphenyl-1,3,4-oxadiazole (at *ortho* position) with carbazole (of 9-position) through an aromatic nucleophilic substitution reaction (Tao et al. 2008). The OLED device constructed from this hybrid oxadiazole-carbazole exhibited a high value of external quantum efficiencies, particularly 18.5% in the case of deep red and 20.2% in the case of green electrophosphorescence.

Similarly, in 2010, Tao et al. developed another hybrid compound as 29 (89% yield) utilizing reaction at the meta position of a phenyl ring in triphenylamine (Tao et al. 2010). Tuning the electron transport layer with 1,3,5-tris(N -phenylbenzimidazol-2-yl)benzene, the authors achieved 23% external quantum efficiency and 94.3 lm/W maximum power efficiency in the case of the constructed device. Also, the OLED device exhibited the best electroluminescence

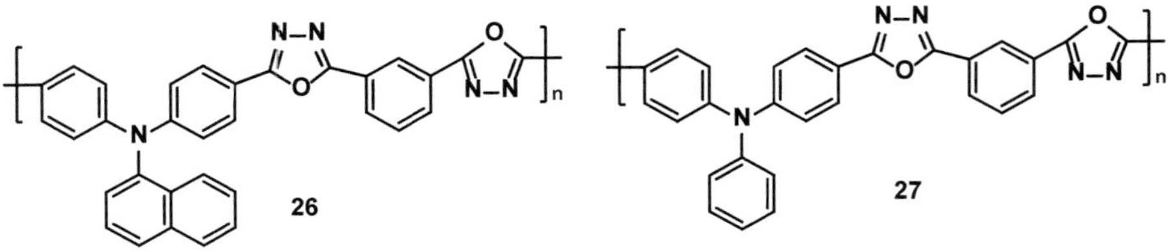

Structures 26 and 27

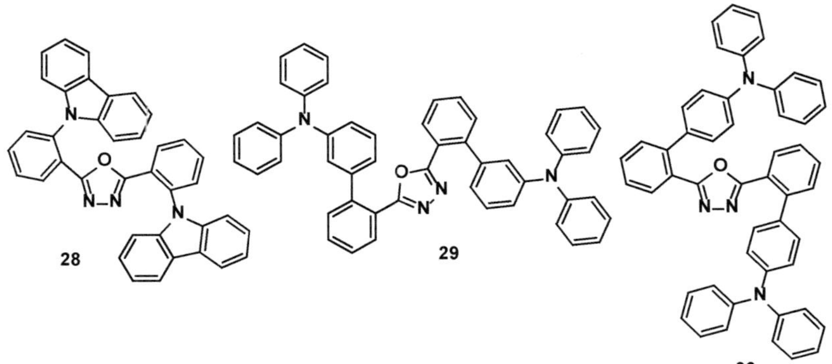

Structures 28–30

(105 lm/W as maximum power efficiency), while a thin layer of compound 29 was introduced between the emissive layer and hole transport. Therefore, the device with 29 could be utilized for designing excellent phosphorescent OLEDs.

Tao et al. also studied the isomer of compound 29 as 30 for phosphorescent OLED applications (Tao et al. 2010). Compound 30 exhibited a high LUMO level (5.25 eV). The efficiency of phosphorescent OLED improved significantly with adding 1,3,5-tris(N-phenylbenzimidazol-2-yl)benzene as an electron transport layer. Power efficiency, peak current efficiency, and external quantum efficiency increased to 97.7 lmW^{-1}, 90.0 cdA^{-1}, and 23.5%, respectively, on incorporating 10 nm of compound 30 as an exciton blocking layer.

In order to develop non-doped OLED devices with blue emitting characteristics, Tao et al. prepared bipolar materials with compounds 31 and 32 utilizing Suzuki cross-coupling reaction catalyzed by palladium(0) (Tao et al. 2009). Compounds 31 and 32 exhibited 2.35 and 2.46 eV triplet energy, respectively. Better localization

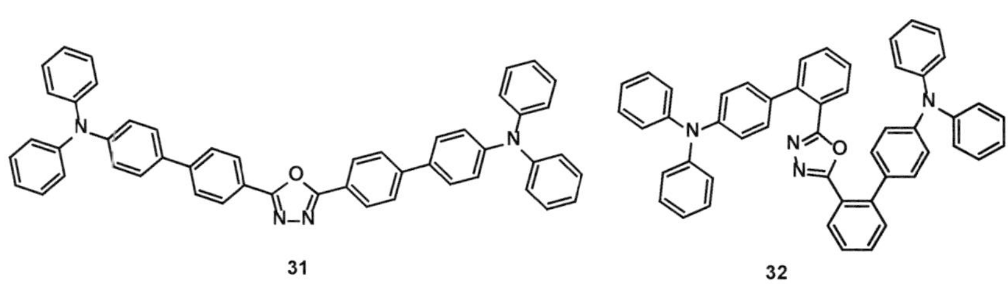

Structures 31 and 32

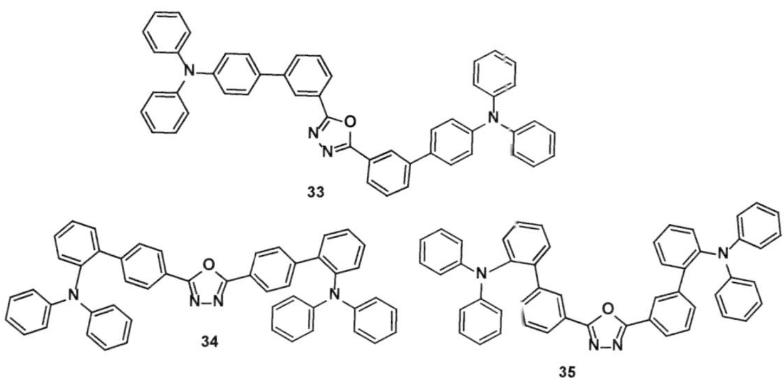

Structures 33–35

of charges and a bigger triplet energy gap in compound 32 led to superior device performance.

The performance of OLED devices can be fine-tuned with suitable structural arrangements to yield better photophysical and optical properties. Tao et al. also studied oxadiazole-triphenylamine-based compounds 33–35 for designing deep-red phosphorescent materials (Tao et al. 2010). Compounds 34 and 35 with attached ortho-triphenylamine exhibited blue-shifted emission, small intramolecular charge transfer, large energy gap, and elevated triplet energy compared to compound 33, where triphenylamine is connected through para position. OLED device fabricated with compound 35 exhibited 21.6% EQE_{max} with deep-red phosphorescence response.

Augmenting molecular size can improve morphological, photophysical, and electrochemical characteristics. In 2012, Gong et al. tailored a series of hybrid oxadiazole-arylamine-based compounds 36–39 through a double silicon bridge (Gong et al. 2012). Higher glass-transition temperature (92–190°C), remarkable solubility, better solution processability, and high thermal stability are some of the features that boost the performance of the tailored compounds. Engineering through tetra-meta-position and incorporation of diphenylamine scaffold led to a higher HOMO level (5.30 eV) as well as very high triplet energy (2.72 eV) in the case of compound 38. Blue OLED device developed from compound 38 resulted in 10.7% EQE_{max}, 23.4 cdA^{-1} as maximum current efficiency, and 10.2 lmW^{-1} as maximum power efficiency. Reddy et al. synthesized a series of compounds 40–45 containing anthracene emitter and oxadiazole units in five synthetic steps from commercially available anthracene and tert-Butyl alcohol (Reddy et al. 2009). OLED materials utilizing compounds 40–45 exhibited a rise in thermal robustness and significant glass-transition temperature (52–131°C). Both CH – π and π–π interactions are mainly responsible for stabilizing the synthesized compounds 40 and 45 in the solid state. These compounds could be useful as electron transport layers in multifunctional OLEDs.

Venkatakrishnan et al. developed twisted bimesitylene centered oxadiazole compounds 46 and 47 in 82% and 87% respectively (Venkatakrishnan et al. 2012). The fabricated OLED devices obtained from compounds 46 and 47 exhibited

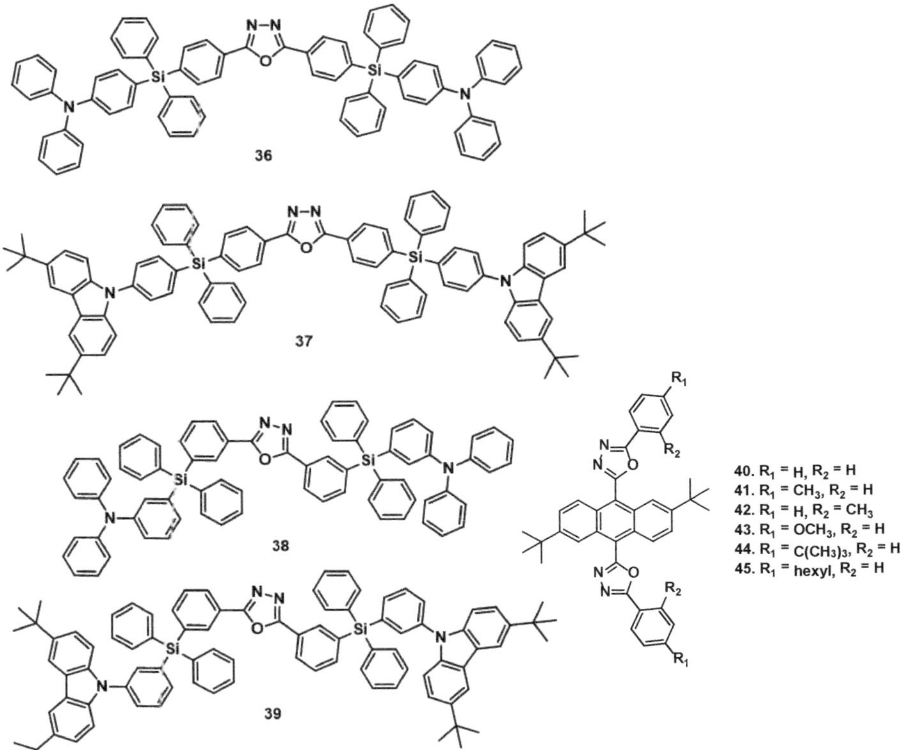

Structures 36–45

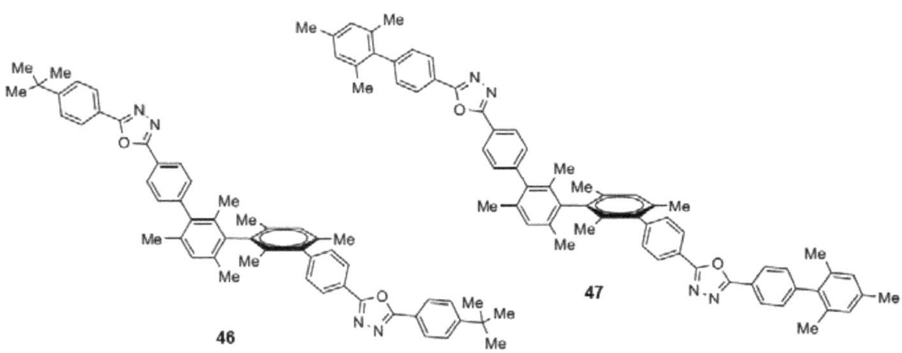

Structures 46 and 47

very high glass-transition temperatures (173°C and 182°C) and a high band gap (ca. 3.71 eV). OLED devices made with these twisted bimesitylenes and $Ir(ppy)_3$ as dopant exhibited 19.0 cdA^{-1} luminance efficiency (Figure 1).

In an effort to enhance triplet energy, Leung et al. at National Taiwan University prepared silane containing oxadiazole materials 48–52 (Leung et al. 2012). The

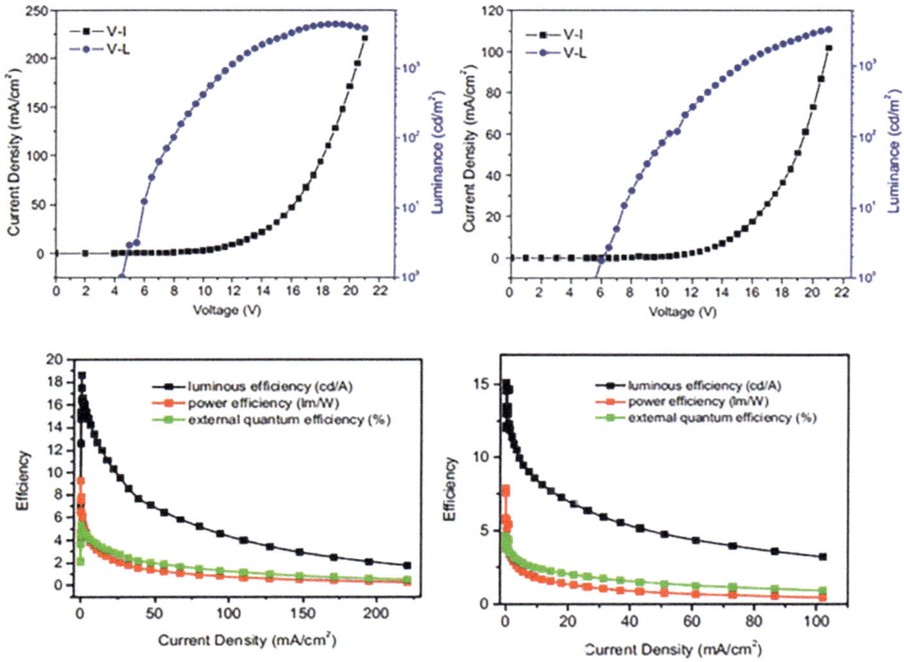

Figure 1. (Top) Plot showing current-voltage-luminance characteristics; (bottom) luminous, power, and external efficiencies vs. current density of the OLED device made from (2,2',2''-(1,3,5-benzenetriyl)-tris(1-phenyl-1-H-benzimidazole) and the synthesized compounds (left) 46 and (right) 47. Reprinted with permission from Elsevier (Venkatakrishnan et al. 2012).

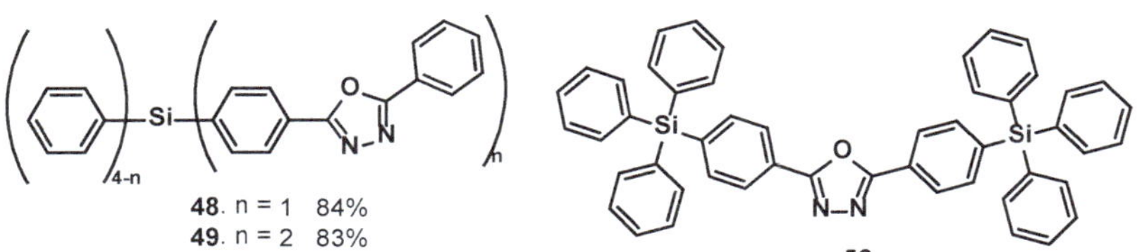

48. n = 1 84%
49. n = 2 83%
50. n = 3 60%
51. n = 4 37%

52

Structures 48–52

introduction of Si atoms inside–aromatic conjugation facilitated high triplet energy as well as large band gap. Compound 52 with symmetrically positioned oxadiazole motif exhibited 5,124 cd/m^2 as maximum luminance, 6.9 V as turn-on voltage, 13.1% as external quantum efficiency, and 39.9 cd/A as current efficiency, indicating novel host material for phosphorescent OLED.

In order to develop cheaper solution-processable host materials, Gong et al. isolated oxadiazole compound 53 with a double silicon bridge between arylamine and oxadiazole as a white solid in 62% yield (Gong et al. 2013). This extended designed compound 53 demonstrated greater thermal stability and high glass-transition temperature (149°C). Subsequently, an OLED device with extended molecular size

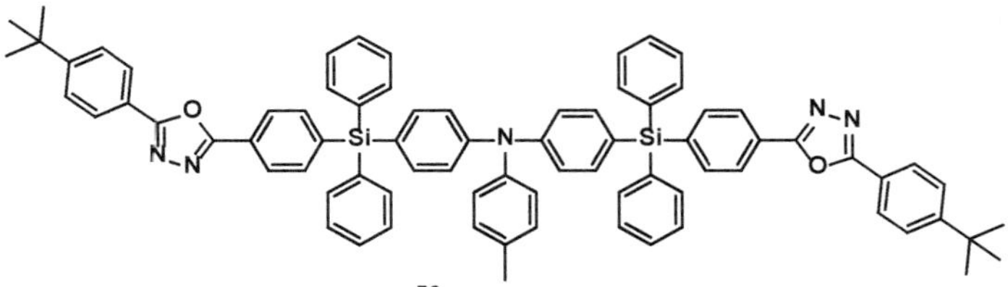

53

Structure 53

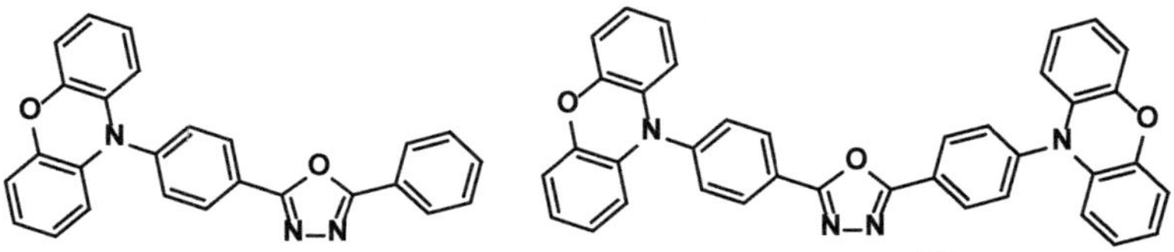

54 55

Structures 54 and 55

exhibited 15.3 cd/A as maximum current efficiency, which is suitable for designing efficient blue phosphorescent materials.

In 2013, researchers at Kyushu University developed donor-acceptor-donor-type efficient TADF materials with oxadiazole analogs 54 and 55 (Lee et al. 2013). Compounds 54 and 55 displayed green emission response. Compound 55 exhibited a very high photoluminescence quantum yield (87%) when doped with bis(2-(diphenylphosphino)phenyl)ether oxide. OLED devices prepared from compound 55 yielded 14.9% EQE_{max}.

In 2013, Chang et al. observed that introducing CsF/Al while developing oxadiazole (56–57) based phosphorescent OLED enhanced electron transport and simultaneously decreased interface resistance (Chang et al. 2013). The authors observed very high current efficiency (17.3 cd/A) and 8.86 lm/W maximum luminous efficiency due to effective n-type doping. By introducing oxadiazole at the C_3 and C_6 positions of carbazole moiety, Chen et al. synthesized a bipolar host material 58

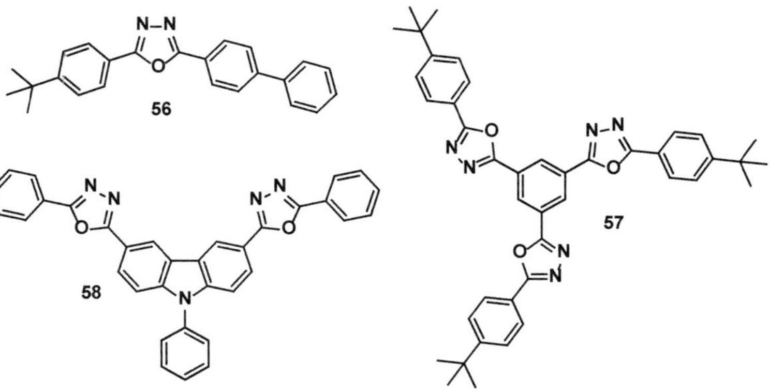

56 57 58

Structures 56–58

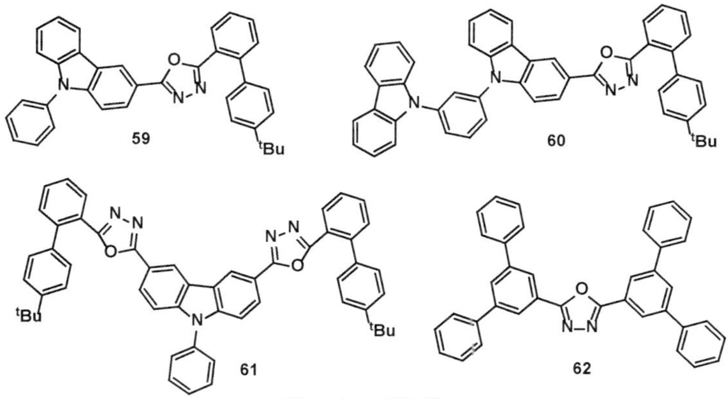

Structures 59–62

in three synthetic steps (Chen et al. 2012). Both HOMO and LUMO orbitals were observed to be delocalized over carbazole and oxadiazole groups in compound 58. Introducing a green phosphor increased EQE_{max} to 17.7%, whereas adding a red phosphor yielded 20.6% EQE_{max}, owing to better delocalization and small triplet energy (2.65 eV).

In order to develop robust bipolar host materials, Cheng et al. tailored carbazole-oxadiazole (donor-acceptor) connected compounds 59–61 by varying the D/A ratio (Cheng et al. 2013). Tuning the carbazole-oxadiazole ratio influenced several parameters such as energy levels, hole mobility, and thermal as well as morphological stability. Interestingly, the addition of bis(2, N-diphenylbenzimidazolito)iridium(III) acetylacetonate, i.e., $(PBi)_2Ir(acac)$ as emitter while preparing green phosphorescent OLED materials enhanced their performance. Around 20.7%, 20.4%, and 17.3% EQE_{max} were observed from materials containing compounds 59–61, respectively. Carrier transport properties of the compounds 59–61 are shown in Figure 2.

Shih et al. developed a centrally positioned oxadiazole-based material 62 with a symmetrical structure in three synthetic steps (70% yield) (Shih et al. 2015). The material displayed a high T_g (thermal stability) value (220°C) as well as high triplet energy (2.76 eV). Several advantages of the material were observed, such as successful enhancement of charge balance, electron transport, blocking holes and excitons, etc. It was successfully utilized as a universal electron transporter for green, blue, and red phosphorescent tools. The green phosphorescent device displayed 97.6 cd A^{-1} as a high value of current efficiency.

In 2016, Chidirala et al. synthesized pyrene-oxadiazole scaffolds 63–65 for green-emitting application materials (Chidirala et al. 2016). Incorporation of 2,3,5,6-tetrafluoro-7,7,8,8-tetracyanoquinodimethane as a hole-injecting layer and bathocuproine as a well-known electron transport material during OLED device fabrication yielded 9.56 lm/W power efficiency, 13.56 cdA^{-1} current efficiency along with 5.34% quantum yield. The study might be useful for designing efficient electron transporters using hybrid chemical entities.

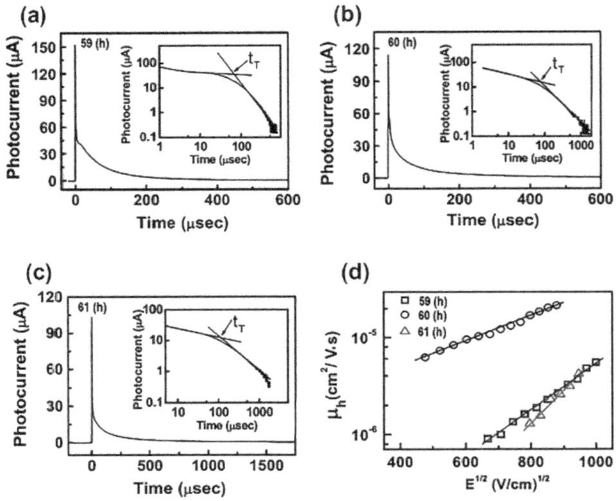

Figure 2. Graphs showing hole transient photocurrent signals for (a) compound 59 (1.8 μm thick) at $E = 8.3 \times 10^5$ V/cm, (b) compound 60 (2.2 μm thick) at $E = 4.1 \times 10^5$ V/cm, and (c) compound 61 (1.9 μm thick) at $E = 8 \times 10^5$ V/cm. Insets are the double logarithmic plots of (a–c). (d) Hole mobilities of compounds 59–61 plotted concerning $E^{1/2}$. Reprinted with permission from Elsevier (Cheng et al. 2013).

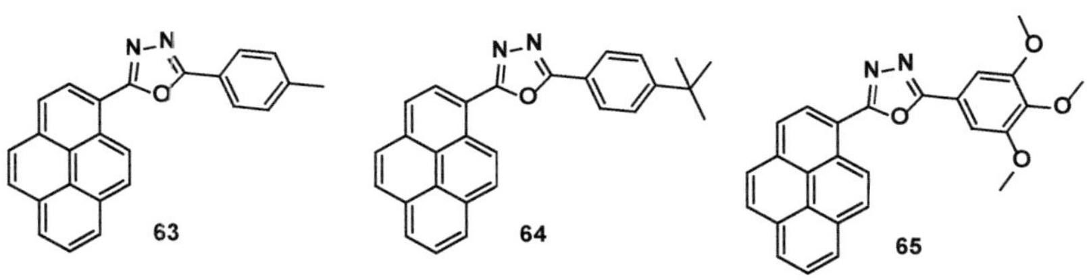

Structures 63–65

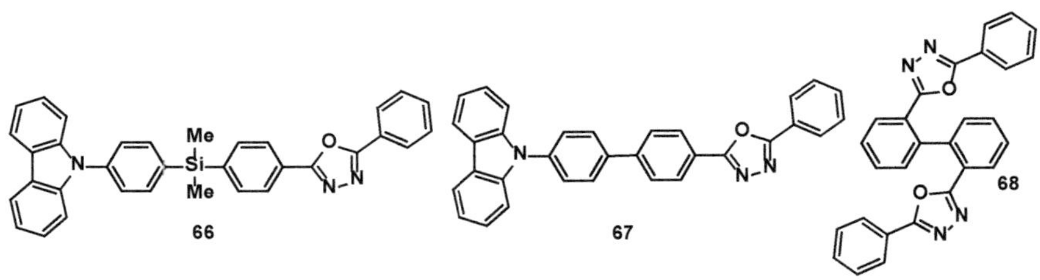

Structures 66–68

Introducing tetrahedral silicon structural motifs in OLED materials can significantly improve band gap and triplet energy levels (Han et al. 2009). Silicon atoms enhance device performance by breaking conjugation character while maintaining thermal stability (Yi et al. 2015). Lee et al. synthesized compounds 66 and 67 with or without tetrahedral silicon scaffold (Lee et al. 2016). Around 15.83% EQE_{max} and 54.55 lm/W maximum power efficiency were observed for OLED devices containing compound 66. The incorporation of tetrahedral silicon moiety also improved the HOMO-LUMO band gap. In another study in 2007, Leung

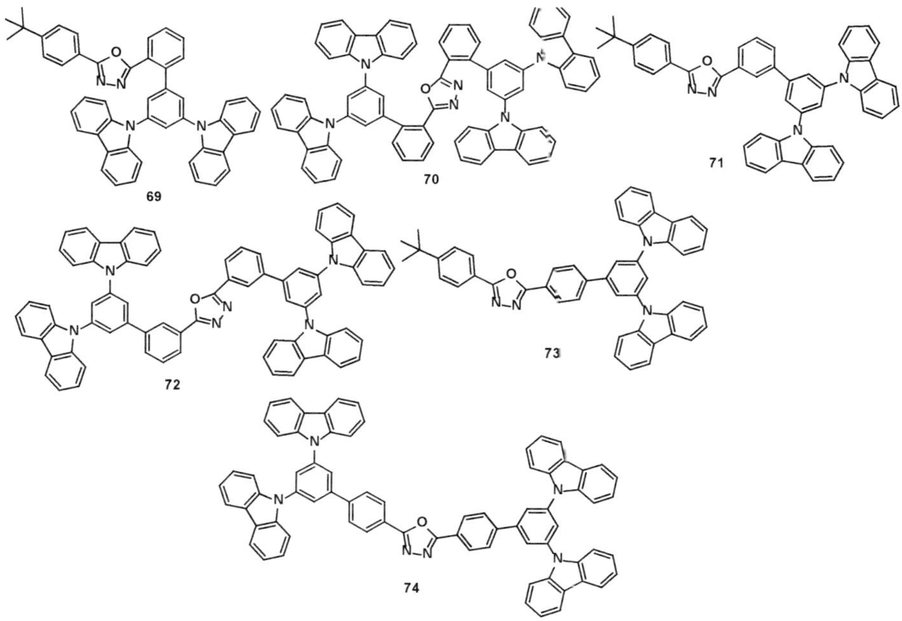

69 70 71 72 73 74

Structures 69–74

et al. observed 26 cd/A efficiency and 43,000 cd/m^2 maximum brightness of the phosphorescent OLED device while incorporating compound 68 as host material (for electron transport) with $Ir(ppy)_3$ [Tris(2-phenylpyridine)iridium] (Leung et al. 2007).

Zhao et al. studied the influence of the carbazole-oxadiazole (donor-acceptor) ratio on phosphorescent OLED utilizing compounds 69–74 that are made up of 1,3-Bis(carbazole-9-yl)benzene as hole transport host and oxadiazole derivatives through a benzene linkage (Zhao et al. 2017). The authors prepared different types (blue, green, and red) of phosphorescent OLEDs and observed interesting patterns concerning ortho/para linkage. Compounds such as 69 and 70 fixed through ortho linkage structurally produced maximum triplet energy value (2.8 eV) and exhibited blue-emitting phosphorescent behavior (Figure 3). Similarly, compounds such as 73 and 74 that are connected through para linkage displayed maximum current densities with short triplet energy values.

Wong et al. developed a series of four TADF emitters 75–78 by fixing two carbazole units at the ortho position of the benzene derivative (Wong et al. 2018). The addition of hosts such as 1,3-bis(carbazol-9-yl)benzene and 2,8-bis(diphenylphosphoryl) dibenzo[b,d] thiophene during device construction resulted in 14–55% photoluminescence quantum yields. Devices prepared from compound 77 yielded 11.2% EQE_{max} with sky-blue emission, whereas devices constructed from compound 78 produced 6.6% EQE_{max} with deep-blue emission. Similarly, CIE color coordinates for OLED devices made from compounds 77 and 78 were (0.17, 0.25) and (0.15, 0.11), respectively.

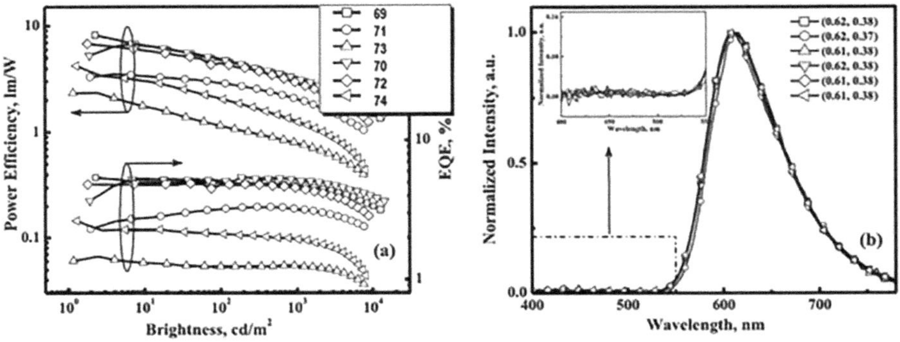

Figure 3. (Left) Comparison of power efficiency and external quantum efficiency versus current density characteristics; (right) electroluminescence spectra of red phosphorescent OLEDs. The inset shows an enlargement of emission from 400–500 nm. Reprinted with permission from Elsevier (Zhao et al. 2017).

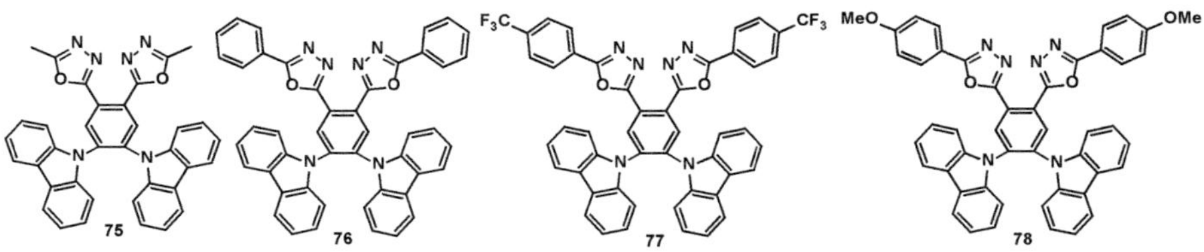

75 76 77 78

Structures 75–78

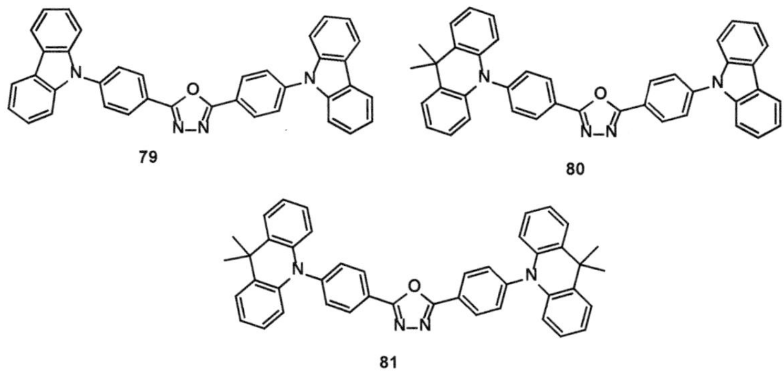

79 80 81

Structures 79–81

Tan et al. developed donor-acceptor-donor type compounds 79–81 in 2019 (Tan et al. 2019). Compounds with substituted acridine, such as 80 and 81, exhibited distinct TADF properties due to small energy splitting (E_{ST}) between singlet and triplet excited states. Acridine-based compounds also displayed very high photoluminescent quantum yields of up to 93% and are potential blue TADF emitters. Similarly, compounds 80 and 81 displayed 14.4% and 22.3% EQE_{max}, respectively.

Li et al. developed a series of oxadiazole-based TADF compounds 82–85 from commercially available 1-bromo-2,4-difluorobenzene precursor (Li et al. 2019). Their emissive color could be easily fine-tuned from sky blue to blue. A small value of singlet-triplet splitting was experienced from the highly twisted conformation

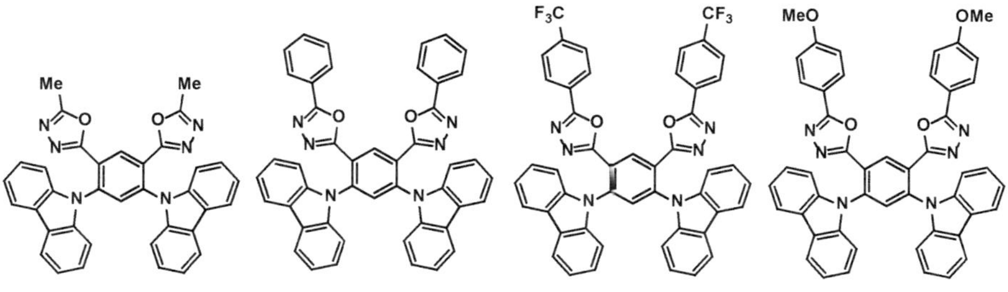

82 83 84 85

Structures 82–85

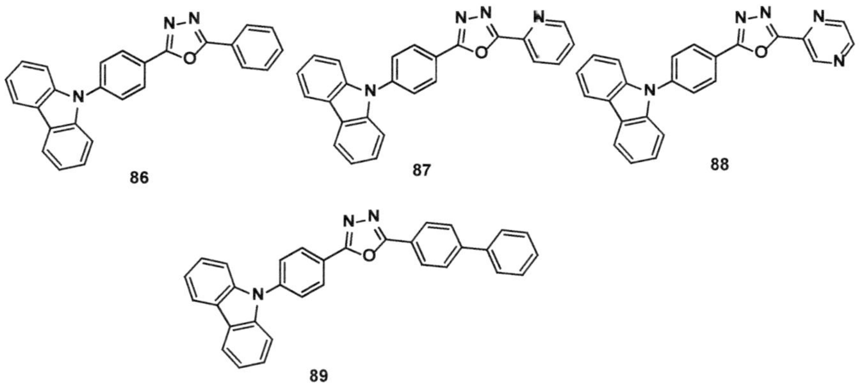

86 87 88 89

Structures 86–89

between oxadiazole and carbazole owing to effective HOMO-LUMO separation. OLED devices prepared from compound 82 exhibited maximum external quantum efficiency (EQE_{max}) as 11.8% with deep-blue emission having color coordinate (CIE) at (0.17, 0.17). However, the device obtained using compound 84 displayed higher EQE_{max} (12.3%) with sky-blue emission having CIE coordinates at (0.17, 0.25).

Tan et al. developed hybrid compounds 86–89 by tuning electronic structures around aromatic nuclei using carbazole and oxadiazole units in three synthetic steps (Tan et al. 2019). Sequential introduction of nitrogen (N) heteroatom in the aromatic structure of the compounds 86–88 resulted in reduced energy gaps and bathochromic shifting of emission bands. All these compounds displayed high photoluminescent quantum yields. The fabricated nanodoped deep blue OLED device (made by the best candidate) prepared by the authors exhibited 4.0% maximum external quantum efficiency and 4,406 cd/m^2 as maximum brightness.

Difluoroboron (BF_2)-attached compounds exhibit better electron withdrawing response and strong luminescence efficiency. Zhou et al. prepared hybrid compounds 90–91 in 2–3 synthetic steps where both phenolic derivative (and its BF_2-analog) and 9,9-dimethylacridine were attached with the help of an oxadiazole moiety (Zhou et al. 2019). Compounds 90–91 exhibited sky-blue emission (maxima in the range of 470490 nm) (Figure 4) both in solution as well as film state, indicating good TADF (thermally activated delayed fluorescence) properties. The developed

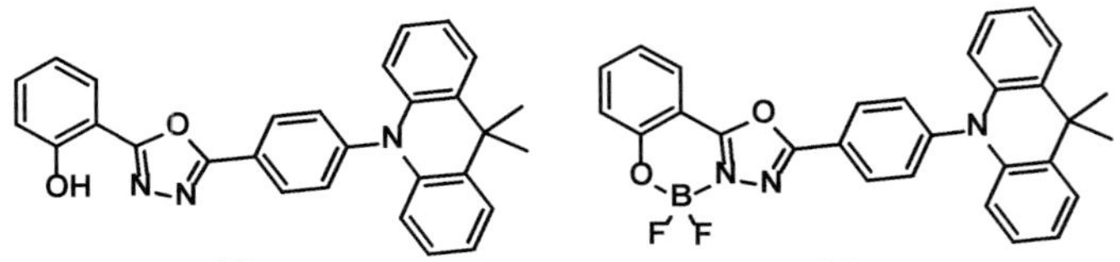

90 91

Structures 90–91

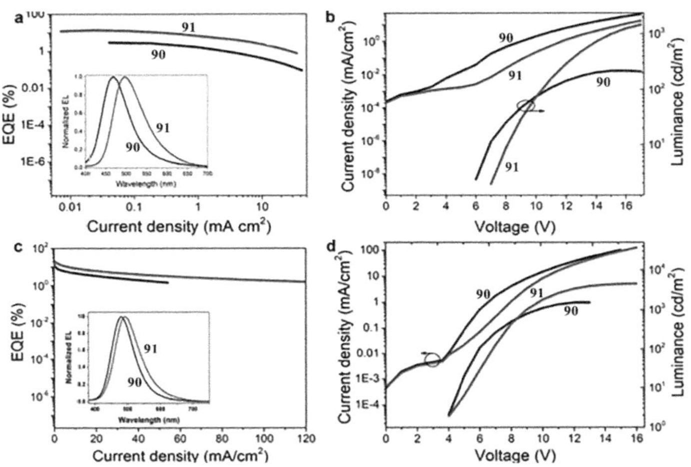

Figure 4. (a) Plot showing EQE vs. current density of OLED devices originating from 90 and 91 (a, c). (b) Luminance-current density-voltage (L–J–V) curve of OLED devices (b, d). Reproduced here with the permission of the American Chemical Society (Zhou et al. 2019).

TADF characteristics are ascribed to the existence of keto form in the triplet state of 90–91. In order to incorporate suitable host material, the authors could tune the external quantum efficiency of the designed OLED materials. In the presence of 9-(4-tert-butylphenyl)-3,6-bis(triphenylsilyl)-9H-carbazole as a host material, the synthesized oxadiazole compounds 90–91 displayed 2.98 and 13.8% maximum external quantum efficiency (EQE), respectively. Interestingly, the EQE parameter for 90–91 increased to 12.1 and 20.1% (Figure 4) in the presence of 10-(4-((4-(9H-carbazol-9-yl)phenyl)sulfonyl)phenyl)-9,9-dimethyl-9,10-dihydroacridine as a host material.

In 2019, researchers at the Georgia Institute of Technology (Zhang et al.) studied host-free OLEDs utilizing oxadiazole-based compound 92 with diagonally connected four carbazole moieties (Zhang et al. 2019). The OLED device prepared from this yellow-green-emitting compound 92 exhibited significant light-emitting properties such as 79 lmW^{-1} as power efficiency, 10 cdm^{-2} as luminance, 73 cdA^{-1} as current efficiency, and 21% of maximum external quantum efficiency. Compound 92 could be utilized as a TADF emitter in organic electronics.

Irradiation of light upon a charge transfer complex at its charge transfer band produces exciplex (also known as excited charge transfer complex) (Guo et al. 2021). Exciplex emissive materials show promising optical properties owing to 100% exciton harvesting, highly advantageous bipolar nature, and preferred dipole

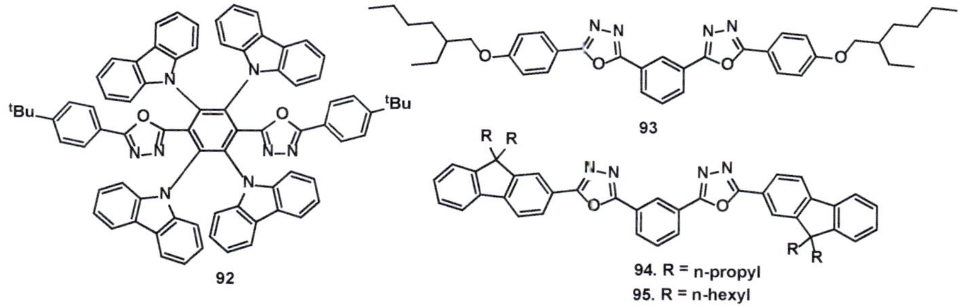

Structures 92–95

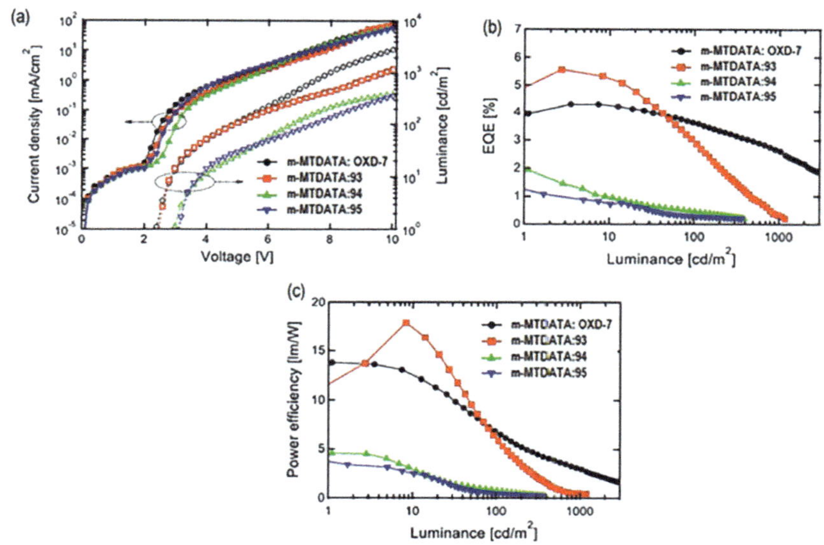

Figure 5. (a) Plot showing voltage vs. current density and luminance, (b) luminance vs. EQE, and (c) luminance vs. power efficiency of OLED devices. Reproduced here with the permission of the American Chemical Society (Saghaei et al. 2021).

alignment (Wang et al. 2019). Oxadiazole derivatives are interesting electron transport materials that could be solution-processed to yield highly efficient exciplex emissive films. In order to utilize exciplex blends in OLED applications, Saghaei et al. tailored three oxadiazole-based compounds 93–95 with better solubility and compared their performance with commercially available electron transport material OXD-7 (Saghaei et al. 2021). Films prepared from these synthesized oxadiazole compounds exhibited greater electron mobilities in comparison to OXD-7. Additionally, these synthesized compounds formed a homogeneous (soluble) solution in the presence of hole-transporting material 4,4',4"-tris[phenyl(m-tolyl) amino]-triphenylamine (m-MTDATA). Blending m-MTDATA with compound 93 displayed 5.3 ± 0.3% as maximum external quantum efficiency (Figure 5) and a high value of photoluminescence quantum yield (21 ± 3%) among three synthesized compounds.

4. Metal Complex-Based Oxadiazole Systems

In an effort to understand the relevance of iridium complexes toward the effectiveness of electroluminescent devices, Tang et al. developed complexes 96 and 97 utilizing 3-(5-(4-(pyridin-2-yl)phenyl)-1,3,4-oxadiazol-2-yl)-9-hexyl-9H-carbazole and its iridium complex (Tang et al. 2011). Complexes 96 and 97 exhibited 188°C and 201°C as high glass-transition temperatures, respectively. Polymer LEDs fabricated from complexes 96 and 97 exhibited very high luminance efficiency of 11.1 and 16.4 cdA^{-1}, respectively (Figure 6).

Oxadiazole-type counter-anions play a significant role in performance enhancement in OLED devices. Taking advantage of the cationic nature of iridium complexes and combining it with oxadiazole-type counter-anions can facilitate energy transfer in the fabricated films. Meng et al. synthesized cationic iridium complexes 98–100 with different counter-anions in 4–5 synthetic steps from commercially available precursors such as Sodium 3-sulfobenzoate and the respective analog (Meng et al. 2020). The authors obtained a 1.4 times improvement in efficiency in

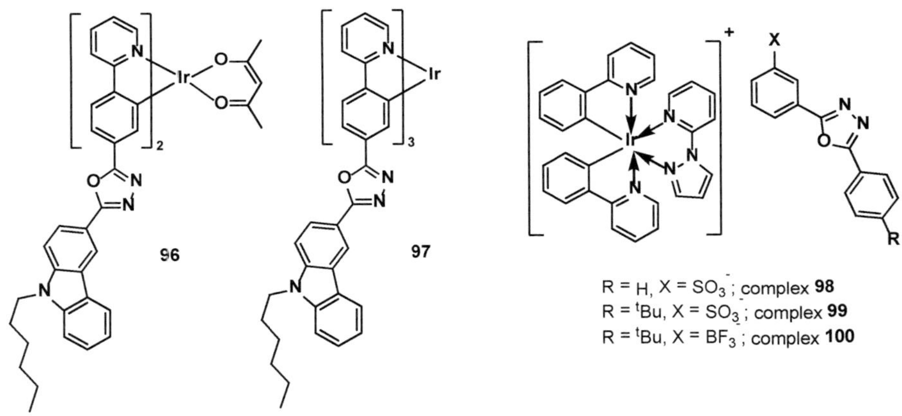

Structures 96–100

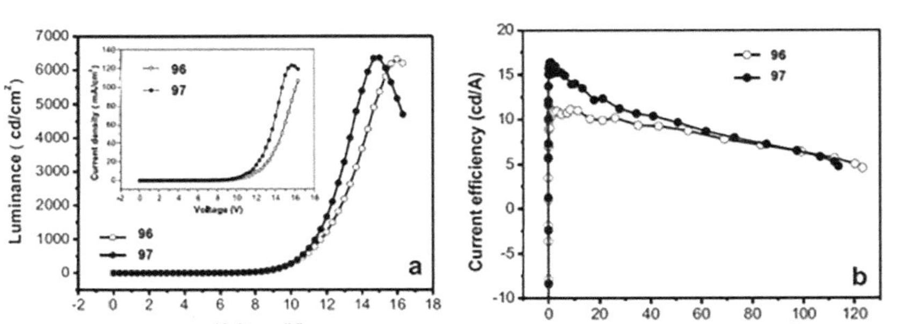

Figure 6. (Left) Plot showing luminance and current density (inset) vs. voltage; (right) plot with current efficiency vs. current density. The graph shows 4 wt % doping concentrations. Reprinted with permission from Elsevier (Tang et al. 2011).

the fabricated solution-processed OLED device prepared from complexes 98–100 compared to the referenced complex with PF_6^- counter anion. Blue-green emitting device prepared from iridium complex with oxadiazole-SO_3^- counter-anion displayed a very high value of EQE_{max} (15.2%) and peak current efficiency (37.6 cdA^{-1}) among available cationic iridium complexes.

5. Polymer-Based Oxadiazole Systems

In 1995, Pei et al. reported synthesizing a set of polymers 101–103 consisting of 1,3,4-oxadiazole units with different conjugation lengths and solubility (Pei and Yang 1995). The non-fluorescent polymer 101 exhibited the largest π–π* bandgap. Interestingly, attachment of extra oxadiazole unit (in case of polymer 102) strengthened its electron transport nature. Alternatively, polymers with longer conjugation lengths, such as 103, exhibited intense blue fluorescence response. The fabricated blue LED device utilizing polymer 103 as the electron injection layer, polyaniline as the hole injection layer (and calcium as a cathode) exhibited 0.1% EQE and 4.5 V voltage. In this way, longer conjugating materials could potentially increase blue fluorescence behavior.

Bao et al. designed and synthesized hybrid polymers 104–106 by attaching oxadiazole units adjacent to the conjugated hole-transporting scaffold such as poly(phenylene vinylene) or poly(phenylene thiophene) (Bao et al. 1998). Polymers 104 and 105 displayed two distinct absorption bands, one at 330 nm from the oxadiazole scaffold and the second at 460 nm. Excitation of 104–106 polymeric thin films at 442 nm exhibited one emission peak at 580 nm in the case of 104–105 and

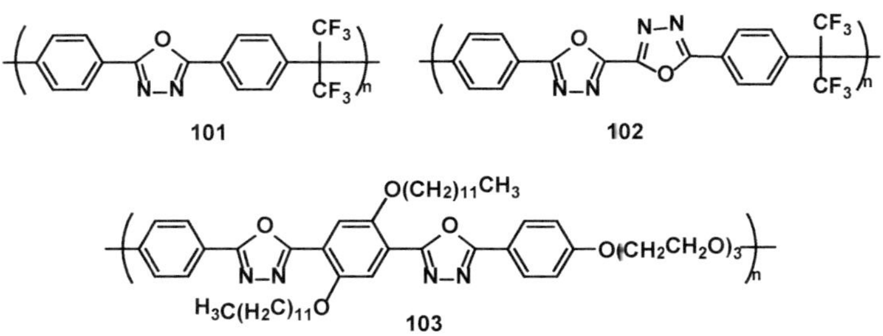

Structures 101–103

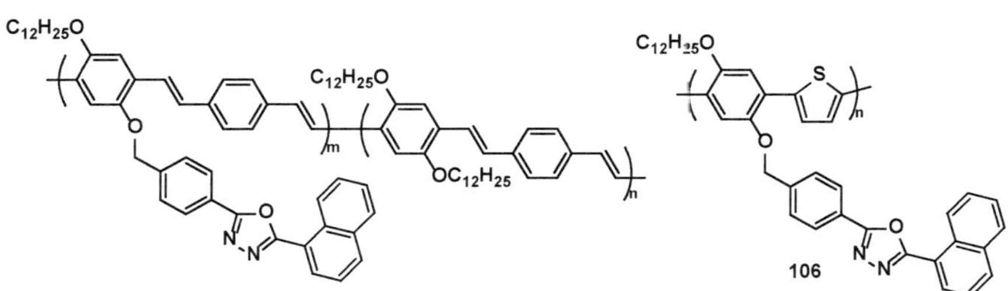

104; when m =1, n = 0
105; when m = 0.5, n = 0.5

Structures 104–106

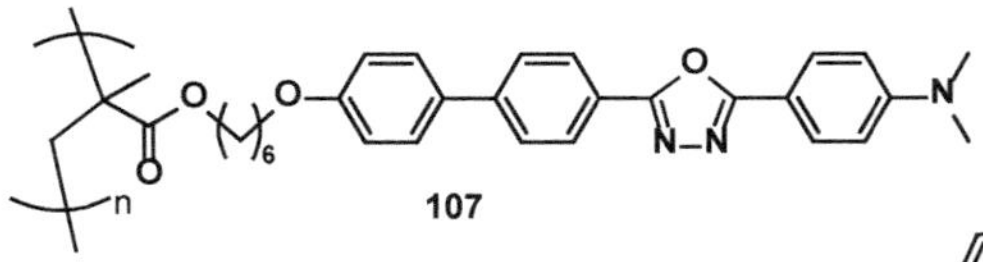

107

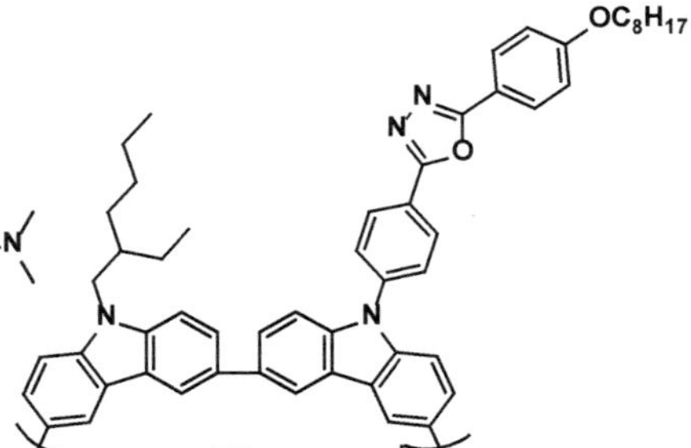

108

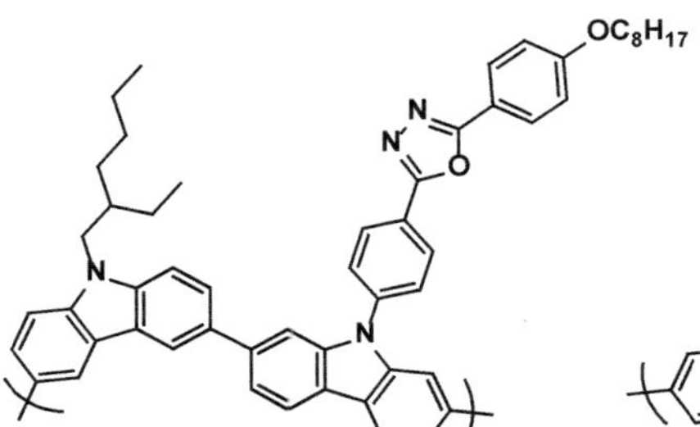

109

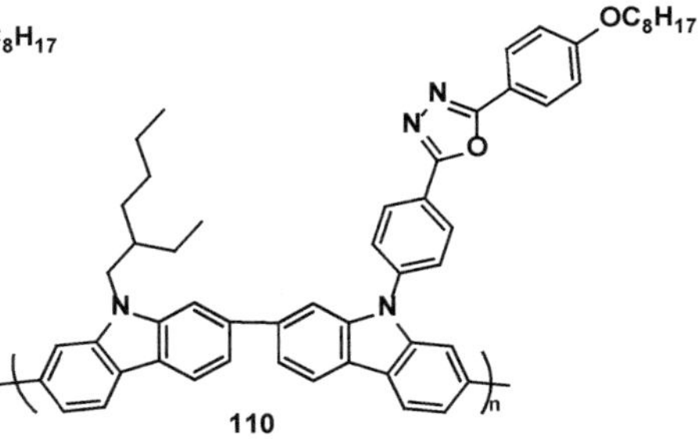

110

Structures 107–110

at 590 nm in the case of 106. Due to hole-transporting, electron-transporting, and light-emitting characteristics in a single species, the designed polymers 104–106 are well suited for fabricating single-layer LED devices.

Polymeric liquid crystals exhibit significant applications in nonlinear optical devices, light-emitting materials, and optoelectronics with interesting phase structures and transitions (Chai et al. 2007; Concellón et al. 2018). Kawamoto et al. developed a polymer liquid crystal utilizing 107, where an oxadiazole scaffold acted as an electron-transporting moiety and the corresponding amine group as a hole-transporting group (Kawamoto et al. 2003). A maximum brightness of around 13 cd/m^2 was noticed with blue emission when polymer 107 was incorporated into the OLED device. In order to tune the triplet energy, Zhang et al. tailored copolymers 108–110 using a carbazole-oxadiazole hybrid system through Suzuki cross-coupling reactions (Zhang et al. 2008). Copolymers were obtained through 3,6-, 3,6-alt-2,7-, and 2,7-linkage at the carbazole moiety. These copolymers displayed good thermal stability and glass-transition temperatures of 211°C, 194°C, and 208°C for 108–110, respectively. Fluorescence quantum efficiencies enhanced remarkably with blue emissions, and the triplet energies were obtained as 2.52, 2.42, and 2.32 eV, respectively, for copolymers 108–110. The fabricated polymeric OLEDs utilizing copolymers 108–110 as host materials and bis(2,4-diphenylquinolinato-N,C$^{2'}$) iridium(acetylacetonate) as guest material exhibited 3.1 to 4.3 V as turn-on voltage.

Sun et al. utilized fluorene-oxadiazole copolymer 111 to develop white light-emitting electrochemical cells (Sun et al. 2010). Copolymer 111 is composed of 25 mol % 5,5'-diphenyl-2,2'-bi-1,3,4-oxadiazole and 75 mol % fluorene.

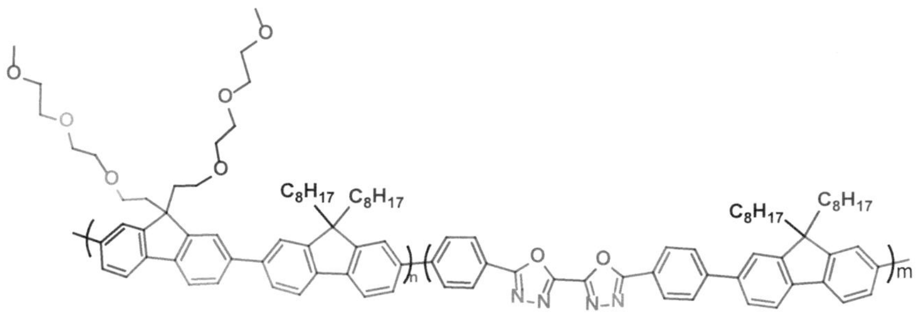

111

Structure 111

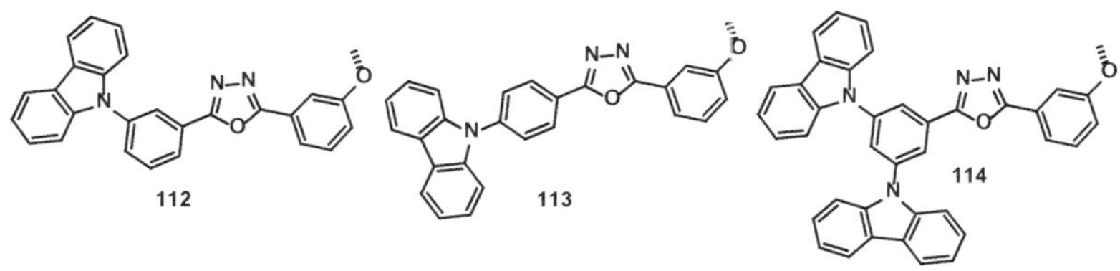

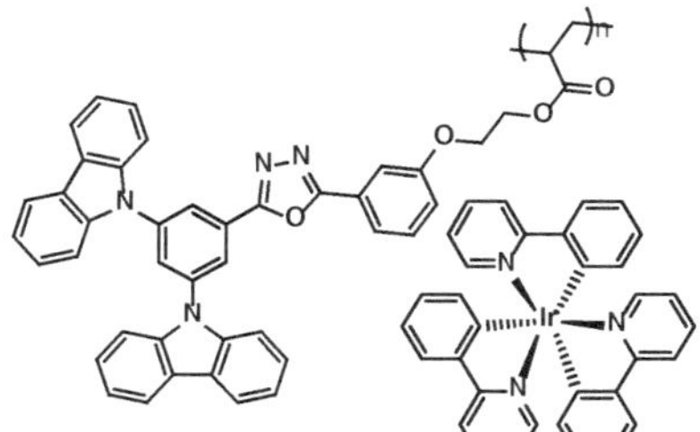

115

Structures 112–115

Attachment of 2-(2-(2-Methoxyethoxy)ethoxy)ethyl group at the C9 position of fluorene facilitated ionic conductivity. About 4.6 V turn-on voltage and 29.4 mA/cm^2 maximum luminance were obtained from the light-emitting electrochemical cell using copolymer 111. White emission was also observed with CIE coordinates (0.24, 0.31).

Zhang et al. developed various polymeric systems 112–115, such as polymethacrylates, polystyrenes, and polynorbornenes via phenylene-linked oxadiazole-carbazole hybrid scaffolds (Zhang et al. 2011). These synthesized polymers displayed good thermal stability and 118–209°C at moderate to high glass-transition temperatures. These polymers could be useful for preparing green and blue-green phosphorescent materials. The fabricated OLED device utilizing polymer 115 displayed 10.0% EQE_{max}, 34.1 cdA^{-1} as current efficiency, and 6.0 V turn-on voltage.

6. Conclusions

Organic and polymeric electronics have achieved tremendous advancement in the last two decades. For example, the quality of electronic and optical devices has significantly increased. Electroluminescent devices made up of oxadiazole moieties exhibit high quantum efficiencies and are suitable for fabricating organic light-emitting diodes. These devices find extensive applications in illumination and display. Well-known electron transporter 1,3,4-oxadiazole, connected to two triphenylamine groups as a hole transporter (as in the case of compound 33), displayed highly emissive OLED characteristics due to discontinuity in conjugation. Water/alcohol-soluble oxadiazole derivatives have been instrumental in fabricating solution-processable devices. Challenges include devising high quantum efficient materials with increased current density. A lower rate of quantum efficiency (with a rise in current density) is associated with loss of charge balance, exciton generation in the hole transport layer, exciton segregation in the presence of an electric field, etc. (Dumur and Goubard 2014). Developing new materials with differing combinations of electron transporter and hole transporter might increase quantum efficiency.

References

Ahn, J., C. Wang, C. Pearson, M. Bryce and M. Petty. 2004. Organic light-emitting diodes based on a blend of poly [2-(2-ethylhexyloxy)-5-methoxy-1, 4-phenylenevinylene] and an electron transporting material. Appl. Phys. Lett. 85(7): 1283–1285.

Bao, Z., Z. Peng, M. E. Galvin and E. A. Chandross. 1998. Novel oxadiazole side chain conjugated polymers as single-layer light-emitting diodes with improved quantum efficiencies. Chem. Mater. 10(5): 1201–1204.

Chai, C.-P., X.-Q. Zhu, P. Wang, M.-Q. Ren, X.-F. Chen, Y.-D. Xu et al. 2007. Synthesis and phase structures of mesogen-jacketed liquid crystalline polymers containing 1, 3, 4-oxadiazole based side chains. Macromol. 40(26): 9361–9370.

Chan, L.-H., R.-H. Lee, C.-F. Hsieh, H.-C. Yeh and C.-T. Chen. 2002. Optimization of high-performance blue organic light-emitting diodes containing tetraphenylsilane molecular glass materials. J. Am. Chem. Soc. 124(22): 6469–6479.

Chang, Y.-T., J.-K. Chang, Y.-T. Lee, P.-S. Wang, J.-L. Wu, C.-C. Hsu et al. 2013. High-efficiency small-molecule-based organic light emitting devices with solution processes and oxadiazole-based electron transport materials. ACS Appl. Mater. Interfaces 5(21): 10614–10622.

Chen, H.-F., L.-C. Chi, W.-Y. Hung, W.-J. Chen, T.-Y. Hwu, Y.-H. Chen et al. 2012. Carbazole and benzimidazole/oxadiazole hybrids as bipolar host materials for sky blue, green, and red PhOLEDs. Org. Electron. 13(11): 2671–2681.

Cheng, S.-H., S.-H. Chou, W.-Y. Hung, H.-W. You, Y.-M. Chen, A. Chaskar et al. 2013. Fine-tuning the balance between carbazole and oxadiazole units in bipolar hosts to realize highly efficient green PhOLEDs. Org. Electron. 14(4): 1086–1093.

Chidirala, S., H. Ulla, A. Valaboju, M. R. Kiran, M. E. Mohanty, M. Satyanarayan et al. 2016. Pyrene–oxadiazoles for organic light-emitting diodes: triplet to singlet energy transfer and role of hole-injection/hole-blocking materials. J. Org. Chem. 81(2): 603–614.

Concellón, A., S. Hernández-Ainsa, J. Barberá, P. Romero, J. L. Serrano and M. Marcos. 2018. Proton conductive ionic liquid crystalline poly (ethyleneimine) polymers functionalized with oxadiazole. RSC Adv. 8(66): 37700–37706.

Dumur, F. and F. Goubard. 2014. Triphenylamines and 1, 3, 4-oxadiazoles: a versatile combination for controlling the charge balance in organic electronics. New J. Chem. 38(6): 2204–2224.

Fukushima, T. 2018. Pi conjugated system bricolage (figuration) toward functional organic molecular systems. Mater. Chem. Front. 2(9): 1594–1594.

Gong, S., Q. Fu, W. Zeng, C. Zhong, C. Yang, D. Ma et al. 2012. Solution-processed double-silicon-bridged oxadiazole/arylamine hosts for high-efficiency blue electrophosphorescence. Chem. Mater. 24(16): 3120–3127.

Gong, S., C. Zhong, Q. Fu, D. Ma, J. Qin and C. Yang. 2013. Extension of molecular structure toward solution-processable hosts for efficient blue phosphorescent organic light-emitting diodes. J. Phys. Chem. C 117(1): 549–555.

Guo, J., Y. Zhen, H. Dong and W. Hu. 2021. Recent progress on organic exciplex materials with different donor-acceptor contacting modes for luminescent applications. J. Mater. Chem. C 9(47): 16843–16858.

Han, W.-S., H.-J. Son, K.-R. Wee, K.-T. Min, S. Kwon, I.-H. Suh et al. 2009. Silicon-based blue phosphorescence host materials: structure and photophysical property relationship on methyl/phenylsilanes adorned with 4-(N-carbazolyl) phenyl groups and optimization of their electroluminescence by peripheral 4-(N-carbazolyl) phenyl numbers. J. Phys. Chem. C 113(45): 19686–19693.

Hughes, G. and M. R. Bryce. 2005. Electron-transporting materials for organic electroluminescent and electrophosphorescent devices. J. Mater. Chem. C 15(1): 94–107.

Kawamoto, M., H. Mochizuki, A. Shishido, O. Tsutsumi, T. Ikeda, B. Lee et al. 2003. Side-chain polymer liquid crystals containing oxadiazole and amine moieties with carrier-transporting abilities for single-layer light-emitting diodes. J. Phys. Chem. B 107(21): 4887–4893.

Lee, A.-R., J. Lee, J. Lee and W.-S. Han. 2016. Silicon-based carbazole and oxadiazole hybrid as a bipolar host material for phosphorescent organic light-emitting diodes. Org. Electron. 38: 222–229.

Lee, D. R., C. W. Lee and J. Y. Lee. 2014. High triplet energy host materials for blue phosphorescent organic light-emitting diodes derived from carbazole modified ortho phenylene. J. Mater. Chem. C 2(35): 7256–7263.

Lee, J., K. Shizu, H. Tanaka, H. Nomura, T. Yasuda and C. Adachi. 2013. Oxadiazole- and triazole-based highly-efficient thermally activated delayed fluorescence emitters for organic light-emitting diodes. J. Mater. Chem. C 1(30): 4599–4604.

Leung, M.-k., C.-C. Yang, J.-H. Lee, H.-H. Tsai, C.-F. Lin, C.-Y. Huang et al. 2007. The unusual electrochemical and photophysical behavior of 2, 2 '-bis (1, 3, 4-oxadiazol-2-yl) biphenyls, effective electron transport hosts for phosphorescent organic light emitting diodes. Org. Lett. 9(2): 235–238.

Leung, M.-k., W.-H. Yang, C.-N. Chuang, J.-H. Lee, C.-F. Lin, M.-K. Wei et al. 2012. 1, 3, 4-Oxadiazole containing silanes as novel hosts for blue phosphorescent organic light emitting diodes. Org. Lett. 14(19): 4986–4989.

Li, Q., L.-S. Cui, C. Zhong, Z.-Q. Jiang and L.-S. Liao. 2014. Asymmetric design of bipolar host materials with novel 1, 2, 4-oxadiazole unit in blue phosphorescent device. Org. Lett. 16(6): 1622–1625.

Li, Q., L.-S. Cui, C. Zhong, X.-D. Yuan, S.-C. Dong, Z.-Q. Jiang et al. 2014. Synthesis of new bipolar host materials based on 1, 2, 4-oxadiazole for blue phosphorescent OLEDs. Dyes Pigms. 101: 142–149.

Li, Z., W. Li, C. Keum, E. Archer, B. Zhao, A. M. Slawin et al. 2019. 1, 3, 4-Oxadiazole-based deep blue thermally activated delayed fluorescence emitters for organic light emitting diodes. J. Phys. Chem. C 123(40): 24772–24785.

Liou, G.-S., S.-H. Hsiao, W.-C. Chen and H.-J. Yen. 2006. A new class of high T g and organosoluble aromatic poly (amine−1, 3, 4-oxadiazole) s containing donor and acceptor moieties for blue-light-emitting materials. Macromol. 39(18): 6036–6045.

Meng, X., P. Wang, R. Bai and L. He. 2020. Blue-green-emitting cationic iridium complexes with oxadiazole-type counter-anions and their use for highly efficient solution-processed organic light-emitting diodes. J. Mater. Chem. C 8(18): 6236–6244.

Oyston, S., C. Wang, G. Hughes, A. S. Batsanov, I. F. Perepichka, M. R. Bryce et al. 2005. New 2, 5-diaryl-1, 3, 4-oxadiazole–fluorene hybrids as electron transporting materials for blended-layer organic light emitting diodes. J. Mater. Chem. 15(1): 194–203.

Paun, A., N. Hadade, C. Paraschivescu and M. Matache. 2016. 1, 3, 4-Oxadiazoles as luminescent materials for organic light emitting diodes via cross-coupling reactions. J. Mater. Chem. C 4(37): 8596–8610.

Pei, Q. and Y. Yang. 1995. 1, 3, 4-Oxadiazole-containing polymers as electron-injection and blue electroluminescent materials in polymer light-emitting diodes. Chem. Mater. 7(8): 1568–1575.

Reddy, M. A., A. Thomas, K. Srinivas, V. J. Rao, K. Bhanuprakash, B. Sridhar et al. 2009. Synthesis and characterization of 9, 10-bis (2-phenyl-1, 3, 4-oxadiazole) derivatives of anthracene: Efficient n-type emitter for organic light-emitting diodes. J. Mater. Chem. 19(34): 6172–6184.

Saghaei, J., T. Leitner, V. T. N. Mai, C. S. K. Ranasinghe, P. L. Burn, I. R. Gentle et al. 2021. Emissive material optimization for solution-processed Exciplex OLEDs. ACS Appl. Electron. Mater. 3(11): 4757–4767.

Shih, C.-H., P. Rajamalli, C.-A. Wu, M.-J. Chiu, L.-K. Chu and C.-H. Cheng. 2015. A high triplet energy, high thermal stability oxadiazole derivative as the electron transporter for highly efficient red, green and blue phosphorescent OLEDs. J. Mater. Chem. C 3(7): 1491–1496.

Song, L., Y. Hu, Z. Liu, Y. Lv, X. Guo and X. Liu. 2017. Harvesting triplet excitons with exciplex thermally activated delayed fluorescence emitters toward high performance heterostructured organic light-emitting field effect transistors. ACS Appl. Mater. Interfaces 9(3): 2711–2719.

Sun, M., C. Zhong, F. Li, Y. Cao and Q. Pei. 2010. A fluorene−oxadiazole copolymer for white light-emitting electrochemical cells. Macromol. 43(4): 1714–1718.

Tamoto, N., C. Adachi and K. Nagai. 1997. Electroluminescence of 1, 3, 4-oxadiazole and triphenylamine-containing molecules as an emitter in organic multilayer light emitting diodes. Chem. Mater. 9(5): 1077–1085.

Tan, Y., B. Rui, J. Li, Z. Zhao, Z. Liu, Z. Bian et al. 2019. Blue thermally activated delayed fluorescence emitters based on a constructing strategy with diversed donors and oxadiazole acceptor and their efficient electroluminescent devices. Opt. Mater. 94: 103–112.

Tan, Y., Z. Wang, C. Wei, Z. Liu, Z. Bian and C. Huang. 2019. Nondoped deep-blue fluorescent organic electroluminescent device with CIEy = 0.06 and low efficiency roll-off based on carbazole/oxadiazole derivatives. Org. Electron. 69: 77–84.

Tang, C. W. and S. A. VanSlyke. 1987. Organic electroluminescent diodes. Appl. Phys. Lett. 51(12): 913–915.

Tang, H., Y. Li, C. Wei, B. Chen, W. Yang, H. Wu et al. 2011. Novel yellow phosphorescent iridium complexes containing a carbazole–oxadiazole unit used in polymeric light-emitting diodes. Dyes Pigms. 91(3): 413–421.

Tao, Y., L. Ao, Q. Wang, C. Zhong, C. Yang, J. Qin et al. 2010. Morphological stable spirobifluorene/oxadiazole hybrids as bipolar host materials for efficient green and red electrophosphorescence. Chem. Asian J. 5(2): 278–284.

Tao, Y., Q. Wang, L. Ao, C. Zhong, J. Qin, C. Yang et al. 2010. Molecular design of host materials based on triphenylamine/oxadiazole hybrids for excellent deep-red phosphorescent organic light-emitting diodes. J. Mater. Chem. 20(9): 1759–1765.

Tao, Y., Q. Wang, Y. Shang, C. Yang, L. Ao, J. Qin et al. 2009. Multifunctional bipolar triphenylamine/oxadiazole derivatives: highly efficient blue fluorescence, red phosphorescence host and two-color based white OLEDs. Chem. Commun. (1): 77–79.

Tao, Y., Q. Wang, C. Yang, J. Qin and D. Ma. 2010. Managing charge balance and triplet excitons to achieve high-power-efficiency phosphorescent organic light-emitting diodes. ACS Appl. Mater. Interfaces 2(10): 2813–2818.

Tao, Y., Q. Wang, C. Yang, Q. Wang, Z. Zhang, T. Zou et al. 2008. A simple carbazole/oxadiazole hybrid molecule: an excellent bipolar host for green and red phosphorescent OLEDs. Angew. Chem., Int. Ed. 47(42): 8104–8107.

Tao, Y., Q. Wang, C. Yang, C. Zhong, J. Qin and D. Ma. 2010. Multifunctional triphenylamine/oxadiazole hybrid as host and exciton-blocking material: high efficiency green phosphorescent OLEDs using easily available and common materials. Adv. Funct. Mater. 20(17): 2923–2929.

Tung, Y.-L., P.-C. Wu, C.-S. Liu, Y. Chi, J.-K. Yu, Y.-H. Hu et al. 2004. Highly efficient red phosphorescent osmium (II) complexes for OLED applications. Organometallics 23(15): 3745–3748.

Venkatakrishnan, P., P. Natarajan, J. N. Moorthy, Z. Lin and T. J. Chow. 2012. Twisted bimesitylene-based oxadiazoles as novel host materials for phosphorescent OLEDs. Tetrahedron 68(36): 7502–7508.

Wang, C., G.-Y. Jung, Y. Hua, C. Pearson, M. R. Bryce, M. C. Petty et al. 2001. An efficient pyridine-and oxadiazole-containing hole-blocking material for organic light-emitting diodes: synthesis, crystal structure, and device performance. Chem. Mater. 13(4): 1167–1173.

Wang, Q., Q.-S. Tian, Y.-L. Zhang, X. Tang and L.-S. Liao. 2019. High-efficiency organic light-emitting diodes with exciplex hosts. J. Mater. Chem. C 7(37): 11329–11360.

Wong, M. Y., S. Krotkus, G. Copley, W. Li, C. Murawski, D. Hall et al. 2018. Deep-blue oxadiazole-containing thermally activated delayed fluorescence emitters for organic light-emitting diodes. ACS Appl. Mater. Interfaces 10(39): 33360–33372.

Wu, F.-I., P.-I. Shih, C.-F. Shu, Y.-L. Tung and Y. Chi. 2005. Highly efficient light-emitting diodes based on fluorene copolymer consisting of triarylamine units in the main chain and oxadiazole pendent groups. Macromol. 38(22): 9028–9036.

Yi, S., J.-H. Kim, W.-R. Bae, J. Lee, W.-S. Han, H.-J. Son et al. 2015. Silicon-based electron-transport materials with high thermal stability and triplet energy for efficient phosphorescent OLEDs. Org. Electron. 27: 126–132.

Yuan, C., S. Saito, C. Camacho, S. Irle, I. Hisaki and S. Yamaguchi. 2013. A π-conjugated system with flexibility and rigidity that shows environment-dependent RGB luminescence. J. Am. Chem. Soc. 135(24): 8842–8845.

Zhang, K., Y. Tao, C. Yang, H. You, Y. Zou, J. Qin et al. 2008. Synthesis and properties of carbazole main chain copolymers with oxadiazole pendant toward bipolar polymer host: tuning the HOMO/LUMO level and triplet energy. Chem. Mater. 20(23): 7324–7331.

Zhang, X., M. W. Cooper, Y. Zhang, C. Fuentes-Hernandez, S. Barlow, S. R. Marder et al. 2019. Host-free yellow-green organic light-emitting diodes with external quantum efficiency over 20% based on a compound exhibiting thermally activated delayed fluorescence. ACS Appl. Mater. Interfaces 11(13): 12693–12698.

Zhang, Y., C. Zuniga, S.-J. Kim, D. Cai, S. Barlow, S. Salman et al. 2011. Polymers with carbazole-oxadiazole side chains as ambipolar hosts for phosphorescent light-emitting diodes. Chem. Mater. 23(17): 4002–4015.

Zhao, Y., Q. Guo, X. Li, Q. Wang and D. Ma. 2017. Influence of the linkage mode and D/A ratio of carbazole/oxadiazole based host materials on phosphorescent organic light-emitting diodes. J. Lumin. 188: 612–619.

Zhou, D., D. Liu, X. Gong, H. Ma, G. Qian, S. Gong et al. 2019. Solution-processed highly efficient bluish-green thermally activated delayed fluorescence emitter bearing an asymmetric oxadiazole–difluoroboron double acceptor. ACS Appl. Mater. Interfaces 11(27): 24339–24348.

Chapter 3

Oxadiazole as Liquid Crystals

1. Introduction

For the first time, an Austrian botanist named Fridrich Reinitzer noticed a twofold melting in cholesteryl benzoate when heated. He discovered that some compounds underwent birefringence (between crossed polarizers) when heated, and upon further heating, it turned into a clear liquid (Reinitzer 1888). These substances later became known as liquid crystals (LCs) by Otto Lehman because they have both solid and liquid qualities. The LCs' anisotropic features can be used in optical displays showing mesophases halfway between solids and liquids (Prakash et al. 2020). Their first generation exhibits that mesophase could only occur in linear (or rod-shaped) molecules with a finite length-to-breadth (L/B) ratio. Later, it was found that molecules do not always need to be rod-shaped in order to display mesophases; instead, non-rod-type molecules can also do so (Chandrasekhar et al. 1977). Then, the chiral LCs with ferroelectric nature and sub-microsecond responsiveness were discovered (Ma et al. 2022). Following this, chiral-containing ferroelectric LCs (FLCs) were used in a variety of technical applications (Gieβelmann and Zugenmaier 1995; Das et al. 2019). The achiral ferroelectric LCs with non-zero polarization in external stimuli were synthesized. Due to their efficient synthesis without expensive chiral components and high yields, the synthesis and characterization of such achiral FLCs have gained impetus in the recent two decades (Niori et al. 1996; Watanabe et al. 1998). The melting of these materials is depicted in the accompanying schematic diagram (Figure 1).

2. Classification of Liquid Crystals

LCs can be categorized based on the change in temperature (thermotropic) or concentration (lyotropic) (Chavda et al. 2022; Chen et al. 2022). Thermotropic LCs display a variety of mesophases as a function of temperature. They are crystalline at low temperatures, whereas above the melting point, they show mesophase behavior. The ordering of different phases can be destroyed by thermal motion at high temperatures, turning them into an isotropic liquid phase (Chavda et al. 2022).

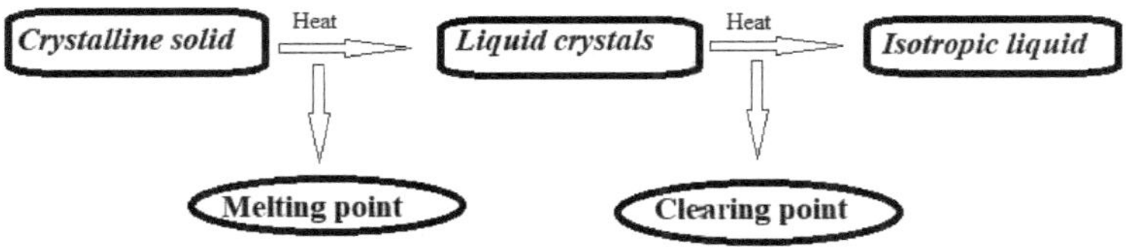

Figure 1. An illustration of the melting of crystalline material.

Based on their shape, thermotropic LCs are further divided into calamatic (Cakar et al. 2022), discotic (Khare et al. 2022), polycatenar (Bais et al. 2019), and bending-shaped (Kumar 2022) LCs. The dissolution of an amphiphile in the solvent results in the formation of lyotropic LCs. The basic structure of these amphiphile compounds consists of a polar head and a non-polar tail. These lyotropic LCs exhibit a variety of phases when the solution concentrations change, and the solvent molecules fill the area around the structures to keep the system fluid (Fairhurst et al. 1998; Chen et al. 2022). The polarity of the solvent, temperature, and concentration all affect the lyotropic phase behavior (Burducea 2004). There may be one or more mesophases present in an LC molecule. Polymesomorphic refers to a compound that displays more than one mesophase. By adjusting the temperature and concentration, the lyotropic exhibits different phase variants and hence is called an Amphotropic LCs. Monotropic occurrence is when an LC compound only displays a phase sequence during the heating/cooling process. Enantiotropic LC phase occurrence is defined as being reversible throughout both the heating and cooling runs of the concentration and dilution processes (Salanger 1999). The classification of liquid crystals is presented in Figure 2.

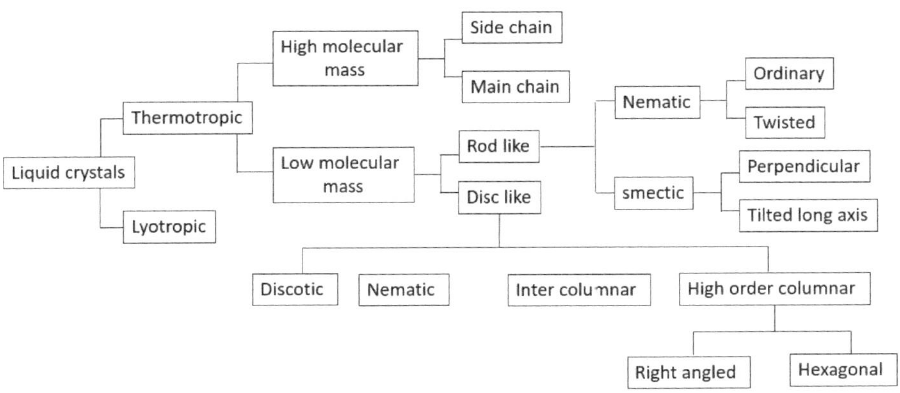

Figure 2. Classifications of liquid crystal.

3. Classification of Liquid Crystalline Phases

Based on their optical textures under crossed polarizers and their X-ray diffraction profiles with temperature fluctuation, various LC phase types can be differentiated (Carsten 1996). When cooling from the isotropic phase to the crystalline phase, these LC phases are distinguished by the molecular orientational order of their constituent molecules (Elston and Sambles 1998). The most typical LC phase is the nematic (N) phase, which is regarded as a high-temperature phase. The molecules lack spatial order but do have long-range orientational order. As a result, the thermotropic phase is the least organized and is, hence, the closest to an isotropic liquid; they are fluid enough for molecules to pass one another easily while maintaining their typical parallel orientations.

Compared to nematics, smectic mesophases exhibit a higher level of order since the molecules are not only aligned in one direction but also further arranged into layers. The degree of order present inside and between layers varies between smectic phases, which exhibit polymorphism. These planes are the only ones where molecules can move. A large number of chemicals can form more than one type of smectic phase. However, because the intralayer interactions between the molecules are stronger than the interlayer forces, the layers are somewhat free to move. Smectics typically have higher viscosities than standard nematics because of the layered structure. The three most significant smectic phases are Smectic A (SmA), Smectic C (SmC), and Chiral Smectic C (SmC), respectively (Gray and Goodby 1984; Khan et al. 2021). Ferroelectric LC phases are found in molecules containing one or more chiral centers, and chirality of the molecules is translated to the chirality of the macroscopic mesophase, forming a helical, chiral assembly. The generated helix's pitch is temperature-dependent, and the helical structure's handedness depends on the presence of a chiral center since one enantiomer produces a left-handed helix while the other produces a right-handed helix (Takanishi 2020; Moriya et al. 2021). Compared to equivalent achiral phases, the symmetry is reduced when chirality is introduced into a mesophase. Chiral mesophases frequently exhibit steric effects that reduce phase stability and lower clearing points (the temperature at which a mesophase transitions to an isotropic phase) (Saupe 1973; Gray and Goodby 1984; Sebastián et al. 2022). Discotic mesogens are disk-shaped, inflexible, flat, symmetrical, aromatic cores with an aliphatic chain perimeter. They can generate either columnar or nematic mesophase. Throughout the sample, the molecules in the discotic nematic phase freely disperse. Columnar phases, in which molecules collect into columns that further organize to produce various two-dimensional columnar assemblages, are the most frequently encountered discotic phases (Chandrasekhar et al. 1977; Percec et al. 2022).

The bent-core (boomerang, bow) mesogens are another name for these banana-shaped LCs. These molecules are bent-core mesogens with an anisotropic form that differs from the classical disk and rod-shaped molecules. Examples are 1,3-disubstituted benzene, 3,5-disubstituted 1,2,4-oxadiazoles, and 2,5-disubstituted 1,3,4-oxadiazole. The first ferroelectric and antiferroelectric LCs without chiral carbon are banana-shaped LCs. When molecules have bent cores, steric packing is preferred because it creates layers with polar axes that are defined by molecules' "bows" pointing in the same direction. This polar order is responsible for the macroscopic chirality of the bulk structure as well as the ferroelectric characteristics of the switchable B2 and B5 crystals. Mesogens in the shape of bananas display a wide range of liquid crystalline states; these compounds display both banana and typical mesophase. It was later established that it had antiferroelectric switching capabilities. These compounds often display the nematic and Bn (B1, B2, ...) phases. The phases will depend on the bond angle between the center rings of the LCs; if the bond angle is lower, it will display nematic phases, and if the bond angle is higher, it will display smectic phases (Sofia et al. 2003; Simić et al. 2022).

4. Oxadiazole as Liquid Crystals

Synonymously called “mesogen,” liquid crystals (LCs) are of great interest to both research and industry because they are synthesized when the properties of two states of matter—solid and liquid confluent (Santos et al. 2021). They have the flexibility of translation between solids and liquids and orientational order, longer shelf life, and are thermodynamically stable (Tschierske 2013). Because of their propensity to align along the axis, LCs have an anisotropy property that is unique for liquids. The loss of orientational order is seen when a molecule has formed anisotropy and such molecules do not undergo a single transition from solid to liquid on heating but rather a series of transitions. The melting point is the first temperature at which a solid enters a phase known as the mesophase. A clearing temperature is when a solid maintains the isotropic phase (Bushby et al. 2013; Han et al. 2018). Due to their unique properties, they are widely recognized as organic semiconductors for the fabrication of efficient, ultra-lightweight, low-power-consuming, low-cost, and flexible components for electrical devices. As a result, new liquid crystal molecules are becoming increasingly popular. Adequate molecular planning is essential to suggest a synthetically feasible structure with the necessary properties. In this context, oxadiazole, the five-membered heterocycles, played a significant role in fabricating more durable materials with various polarities and geometries. Their ease of synthesis tuned physical and chemical, thermal stability, mesomorphic behavior, and high luminescence quantum yield are excellent LCs. This is because their unique structure is equipped with several mesogenic cores, including calamitic, bent cores, polycatenar, and discotic (Ghosh and Lehmann 2017; Kotian et al. 2020; Santos et al. 2021).

It was first discovered in 1888 when Friedrich Reinitzer noticed that cholesteryl benzoate had two melting points in the beginning. At 145°C, the substance was turbid, and at 178.5°C, it turned translucent (Reinitzer 1888). Then, Otto Lehmann used a polarizing microscope to study cholesteryl benzoate in 1889 and concluded that it existed in a new state of matter at this temperature range. He then came up with the phrase “liquid crystal.” The term “mesomorphs,” was proposed by George Friedel. Rinne attempted to rename them as “Paracrystals” in 1933, and Bernal demonstrated in 1951 that naturally occurring LCs are connected to biological processes. Although their behavior is described by the names “anisotropic liquids,” “mesomorphic,” or “mesophases,” the term “LCs” has gained popularity (Kotian et al. 2020).

5. Mesomorphic Features

Oxadiazoles are five-membered heterocycles with one oxygen and two nitrogen atoms. They provide a wide range of biological applications, such as anticancer, anti-inflammatory, antimycobacterial, anticonvulsant, antibacterial, antifungal, hypoglycemic, and antiparasitic drugs (Kotian et al. 2020). The most popular methods for producing 1,3,4-oxadiazole include the oxidation of acyl hydrazones (Nandeesh et al. 2016), thermal or acid-catalyzed cyclization of diacyl hydrazines (Patel et al. 2014), and chloramine-T oxidative cyclization of semicarbazones (Rai et al. 1995).

A new series of 2,2-bis (3,4,5-trialkoxyphenyl)-bi-1,3,4-oxadiazole were synthesized and investigated their mesomorphic behavior. The compound showed Col mesophases at room temperature (Qu and Li 2007). Several non-symmetrical 1,2,4-oxadiazole derivatives' thermal and optical characteristics were extensively studied, showing SmC and N phase behavior. Their mesophase ranges are broad with low melting points (Gallardo et al. 2020). Sung and Lin synthesized a group of compounds with a 1,3,4-oxadiazole moiety containing Me, OMe, Cl, F, CN, and NO_2. Groups and investigated the properties of the liquid crystals. The derivatives displayed a stable SmA, SmE phase. It was observed that the mesomorphic phases, internal quantum efficiencies, optical characteristics, and transition temperature are significantly influenced by terminal substituents (Kotian et al. 2020).

In the presence of pyridine, amidoxime and acyl chlorides can be combined to synthesize 1,2,4-oxadiazole (Chiou and Shine 1989). Oxadiazole-containing bent-shaped LCs were successfully synthesized using NO_2 groups as lateral substituents and C'C groups as connecting groups. The transition temperature decreased with the addition of the terminal NO_2 group, and the Sm phase was discovered to be present (Cristiano et al. 2005). 1,3,4-oxadiazole derivatives were synthesized using thiophene heterocycles, and their mesomorphic characteristics were investigated (Han et al. 2008). The derivatives with the longer alkoxy chain and terminal electron-withdrawing group displayed the SmA phase. They also concluded that terminal groups impact the formation of mesophases. Upon cyclo condensing amidoximes with trifluoroacetic anhydride, 3,5-disubstituted 1,2,4-oxadiazole compounds were synthesized (Guo et al. 2014). The chemical names displayed N and/or Sm phase for different derivatives. By reacting 4-benzyloxybenzonitrile with hydroxylamine hydrochloride, bent-shaped LCs containing 1,2,4-oxadiazole have been synthesized, which produced the appropriate amidoxime and further the desired product (Shanker and Tschierske 2011). Also, 4-[3-(4-benzyloxyphenyl)-1,2,4-oxadiazol-5-yl] phenyl 4-hexyl benzoate was obtained on reacting with 4-(4-n-hexyl benzoyl oxy) benzoic acid which showed an enantiotropic N phase. A new series of 1,3,4-oxadiazole derivatives were synthesized, and their mesomorphic characteristics were investigated. The synthetic materials showed the Colh phase, chirality, OH group, and dependency of the mesomorphic features on the aliphatic chain connected to the rigid disk. The Colh phase of the compound persisted even at ambient temperature (Girotto et al. 2014). Using TG-MS-FTIR, the thermal stability of LCs containing derivatives of 1,3,4-oxadiazole Schiff bases with alkyloxy and acyloxy terminal chains was investigated. The compounds had good thermal stability and could tolerate temperatures up to 330 °C. When compared to acyloxy analogs, it was discovered that compounds with the alkoxy terminal chain had improved thermal stability (Lisa et al. 2011).

Star-shaped 1,3,4-oxadiazole derivatives with alkoxy substitutions were synthesized in conjunction with phenyl ethynylenes. Enantiotropic Col mesophase was present in the discotic molecules that were produced. The compound with a C12 alkyl chain changes into a glassy film with LC characteristics that remained stable for over a year. It is also thought to be a strong contender for displays and non-doped blue-emitting diodes (Prabhu et al. 2012) Phenylene bis-1,3,4-oxadiazole group-containing LCs were synthesized, which showed the characteristics of typical LCs (Chai et al. 2008). Ten novel 1,2,4-oxadiazole bent-rod-shaped compounds were synthesized with polar substituents like nitrogen dioxide (NO_2), hydroxy (OH), amino (NH_2), and iodine (I) on the central core exhibited LC characteristics. Compounds substituted with NO_2, and I lacked mesomorphic properties. Despite having a bent form, compounds showed the Colh phase. It might be due to the hydrogen bonding, steric interactions, and dipoles that result in the dimer, trimer, and tetramer inside a single disk and enable the formation of a Colh phase (Kotian et al. 2020). The 1,3,4-oxadiazole compound, containing gallic acid, was synthesized to study their liquid crystalline characteristics. The compound showed a crystal-to-isotropic transition and exhibited monotropic Colh, which made it evident that the terminal alkyl chain influences how the LC behaves. It was claimed that the produced compounds have improved thermal stability, which makes them suitable materials for organic electronics (Frizon et al. 2014).

In order to study the behavior of 2,5-disubstituted [1,3,4]-oxadiazole in polycatenars 2,5-bis4-[(E)-2-(3,4-diakoxyphenyl) vinyl] phenyl-1,3,4-oxadiazoles were synthesized. Tetracatenar oxadiazoles, dicatenar oxadiazoles, and hexacatenar oxadiazoles all showed lamellar phase, whereas the dicatenar oxadiazoles showed only Col phase. Two phenylenevinylene groups in the tetracatenar caused lamellar phases. As the number of chains increased from dicatenars to tetracatenars to hexacatenars, a drop in the transition temperature was seen (He et al. 2007). Several nonsymmetric compounds with 1,3,4-oxadiazole and biphenyl units and various terminal chains were synthetically produced. The derivatives showed either N or SmA mesophase. It was discovered that the mesophases were caused by the electrical characteristics of terminal groups, such as UV-Vis absorption and photoluminescence (Han et al. 2010). Boomerang-shaped LCs containing 2,5-bis(p-hydroxyphenyl)-1,3,4-oxadiazole showed LC characteristics. P-Dodecyloxyphenyl tails were used to replace the mesogenic core, and the outcome was polymorphism with five phases: N, SmC, SmX, SmY, and SmZ (Dingemans and Samulski 2000). Barberá and colleagues synthesized and studied a unique series of LC materials based on 1,3,5-benzenetrisamide trioxadiazole derivatives with varied side alkoxy chains. They have documented how mesophases depend on various terminal alkoxy chains. Nine alkoxy chains were reported to exhibit the Col phase, while compounds with six alkoxy chains displayed the Colh phase (Barberá et al. 2011).

By treating 3-(4-pyridyl)-5-(4-n-alkoxy) phenyl-1,2,4-oxadiazoles with benzoic acid, 4-chlorobenzoic acid, or 4-methylbenzoic acid, the LCs were created. It was noticed that oxadiazole or carboxylic acid did not exhibit any mesomorphic features. But LC behavior was visible when they were joined together through H-bonding (Parra et al. 2005). Lin and co-workers synthesized a new set of mesogens carrying 1,3,4-oxadiazole and thiophene and displayed the Colh phase (Lin et al. 2015). Tomi and co-workers synthesized and characterized two series of ester compounds based on the two isomers of the oxadiazole ring. On cooling in a specific range, some derivatives of the 1,2,4-oxadiazole isomer showed monotropic N mesomorphism. The 1,3,4-oxadiazole derivatives that were synthesized, however, were discovered to be non-mesomorphic. They concluded that the bending angles of the oxadiazole derivatives substantially influence mesomorphic characteristics (Tomi et al. 2017). Another study investigated the impact of the terminal alkoxy chains in the 1,2,4-oxadiazole core, which had both ester and ether linkages inside the same molecule. They concluded that the length of the terminal alkoxy chain did not affect the character of mesophase (Ali et al. 2017).

Kang and co-workers synthesized bent-shaped molecules with varied side chains. Polymorphism was discovered in oxadiazole derivatives, including calamitic N and SmC phases and the banana Bx phase (Kang et al. 2006). In another study, a series of polycatenar-based LCs containing bis-1,3,4-oxadiazole were investigated to study their liquid crystalline characteristics, which exhibit the Colh phase (Kotian et al. 2020). Based on oxadiazole bis aniline, the synthesis of bent-core LC molecules was reported, and their phase behaviors were investigated. Three distinct alkyl chains were used to create three different molecules. Compounds that have terminal hexyl and terminal octyl chains only display N phases, whereas compounds with terminal dodecyl alkyl chains only display SmC, SmX, and SmY phases, respectively (Kotian et al. 2020).

Saha et al. (2017) created the asymmetrical hockey stick-shaped mesogenic molecules with 1,3,4-oxadiazole as the central core component and a lateral methoxy group. The monotropic N phase was present in compounds with shorter alkyl chains, but the SmA phase was present in the extended derivative. Abboud et al. (2017) explored a similar series with the terminal nitro group and the mesogenic core 1,3,4-oxadiazole. While compounds with longer-terminal alkoxy chains displayed the SmA phase, those with shorter-terminal alkyl chains displayed the N phase. Al-Obaidy et al. (2017) investigated several 1,3,4 and 1,2,4 oxadiazole derivatives with various para-substitutions (CN, $COOCH_3$, OCH_3, NO_2, Cl). They discovered that several groups contributed to the molecule's mesogenity in various ways. SmA, N, and SmC phases could be seen in every derivative that was produced. The mesomorphic properties of 1,3,4- and 1,2,4-oxadiazoles were also studied, and it was discovered that 1,3,4-oxadiazole derivatives lack LC features.

The impact of lateral halogen substitutions in bent-core LCs with oxadiazole rings was described by Nguyen et al. (2015). When the compounds with one halogen and two methyl groups reached room temperature, they displayed the N phase. When halogens were substituted, the N phase persisted and changed to the isotropic phase upon further heating. The synthesis and mesomorphic behavior of tris- and bis-(1,3,4-oxadiazoles) with peripheral quinoxalines were reported by Lin and Lai (2016). The compounds exhibited phase crossover in the temperature range between the rectangular and hexagonal Col phases, which displayed Colr at low temperatures and Colh at high temperatures. They explained it away as being caused by the hefty quinoxaline necklace. A new series of 4-[5-(p-tolyl)-1,2,4-oxadiazol-3-yl] phenyl substituted benzoate derivatives were synthesized by Ali and co-workers and investigated the impact of terminal substituents on LC characteristics (Ali et al. 2019). The para-substituted compounds displayed an enantiotropic N phase over a broad temperature range. In order to study their potential use as dopants in blue phases (BPs), Yu et al. (2019) developed bent-shaped light-responsive azo-oxadiazole derivatives. When compared to straight terminals, they discovered that synthetic derivatives with branching terminals had greater miscibility in the host LC. They also looked at how the rigid core's length affected the BP range and discovered that as it got longer, the rigid core's temperature range likewise grew. The 1,2,4-oxadiazole linked to thiophene was synthesized by Girotto et al. (2016) and showed mesomorphic behavior. They noticed that the increase in the number of chains on the symmetrical and non-symmetrical compounds did not favor the mesomorphic behavior, most likely due to the rise in volume. Mesomorphism in the N phase was also seen in compounds when triple bonds were used as spacers instead of the 1,3,4-oxadiazole in combination. This compound produced a more ordered SmC phase and N phase, and a lower melting temperature was also noted. Furthermore, 1,3,4-oxadiazoles derivatives were synthesized by Kuo et al. (2016), who also studied their mesomorphic behavior. The limited mesophase temperature range and higher clearing temperatures were present in all the synthetically produced derivatives' N or/and Sm phases. In one of their schemes, Choi et al. (2011) reported synthesizing a new series of 1,3,4-oxadiazole derivatives and investigated their LC characteristics. They also detailed the synthesis of 1,3,4-oxadiazole derivatives with lollipop shapes. Derivatives that combined rod- and disk-shaped mesogen produced monotropic discotic Col mesophase.

Oxadiazoles are one of the most widely used heterocycles in LCs among the five-membered heterocycles. Due to the simplicity of synthesis, there are numerous papers on the synthesis and mesomorphic investigations of 1,3,4- and 1,2,4-oxadiazoles. When mesomorphic studies are considered, it is discovered that 1,2,4-oxadiazoles exhibit better mesomorphic features, indicating the importance of the bending angle. When joined to substituted phenyl rings and another five-membered heterocyclic ring, 1,3,4-oxadiazole rings display mesomorphism. Mesophase types include banana mesophases, discotic or Col mesophases, and traditional calamitic mesophases (N, SmA, and SmC).

6. Conclusion

The applications of LCs, which are intriguing materials, are used in a variety of scientific fields, including physics, chemistry, arithmetic, biology, and technology. They are frequently used in the fabrication of devices for various applications. Due to its asymmetric electron cloud distribution, the 1,2,4-oxadiazole is a promising core to display "B" phases. The 1,2,3, 1,2,4, 1,2,5, and 1,3,4-oxadiazoles exhibit remarkable biological and physiological properties. The oxadiazole derivatives have also been noted to be effective OLEDs, photovoltaic, and EL devices. It has been widely acknowledged that there has been a significant advancement in the structure-property connections of liquid crystalline 1,3,4-oxadiazole derivatives. The 1,3,4-oxadiazole unit, one of the most significant synthons, has been extensively employed to create numerous LCs with different mesophases, such as uniaxial nematic, smectic, columnar phases and even the elusive nematic phase with biaxiality. But there are still a lot of scientific questions in this area that require inquiry, and research needs thorough investigations. Last but not least, even though numerous liquid crystalline 1,3,4-oxadiazole derivatives have been developed, most of them are used for theoretical and basic research. Future research into the practical application of novel materials must be considerably expanded.

References

Abboud, H. J., S. J. Lafta and I. H. R. Tomi. 2017. Synthesis and characterization of asymmetrical mesogenic materials based on 2,5-disubstituted-1,3,4-oxadiazole. Liq. Cryst. 44(14-15): 2230–2246

Ali, G. Q., I. H. R. Tomi and J. Dispers. 2019. Study the effect of different terminal substituents on the liquid crystalline properties of the new synthesized 1,2,4-oxadiazole derivatives. Sci. Technol. 40(8): 1093–1100.

Ali, G. Q. and I. H. R. Tomi. 2018. Synthesis and characterization of new mesogenic esters derived from 1,2,4-oxadiazole and study the effect of alkoxy chain length in their liquid crystalline properties. Liq. Cryst. 45(3): 421–430.

Al-Obaidy, M. M. A. R., I. H. R. Tomi and H. J. Jaffer. 2017. Non-symmetrically (1,2,4- and 1,3,4-) oxadiazole homologous: synthesis, characterisation and study the effect of different substituents on their mesophase behaviours. Liq. Cryst. 44(7): 1131–1145.

Bais, A., Z. Ashraf, M. N. Tahir, F. Perveen, M. Abbas and I. Ahmed. 2019. Synthesis, single crystal X-ray structure and thermal analysis of a novel polycatenar liquid crystal: Theoretical and experimental approaches. J. Mol. Str. 1177: 1–8.

Barberá, J., M. A. Godoy, P. I. Hidalgo, M. L. Parra, J. A. Ulloa and J. M. Vergara. 2011. Columnar liquid crystalline benzenetrisamides with pendant 1, 3, 4-oxadiazole groups. Liq. Cryst. 38(6): 679–688.

Burducea, G. 2004. Lyotropic liquid crystals I. specific structures. Romanian Reports in Physics 56(1): 66–86

Bushby, R. J., S. M. Kelly and M. O'neill (eds.). 2013. Liquid Crystalline Semiconductors - Materials, Properties and Applications. Springer Series in Materials Science (SSMATERIALS: 169), Springer: The Netherlands.

Cakar, A. E., F. Cakar, H. Ocak, S. Karavelioglu, B. B. Eran and O. Cankurtaran. 2022. Determination of the surface thermodynamic characteristics and the structural isomer separation ability of new synthesized phenylbenzoate-based three-ring calamitic liquid crystals by inverse gas chromatography. J. Mol. Str. 1265: 133379.

Carsten, T. 1996. Molecular self-organization of amphotropic liquid crystals. Prog. Polym. Sci. 21: 775–852. Amphotropic liquid crystals. Current Opinion in Colloid and Interface Science 2002, 7: 355–370

Chai, C., Q. Yang, X. Fan, X. Chen, Z. Shen and Q. Zhou. 2008. Synthesis, characterisation and liquid crystal properties of 2,5-bis[5-alkyl(alkoxy)phenyl-1,3,4oxadiazole]bromobenzenes. Liq. Cryst. 35(2): 133–141.

Chandrasekhar, B. K., Sadashiva and K. A. Suresh. 1977. Liquid crystals of disc-like molecules. Pramana- J. of Phys. 9: 471–480.

Chavda, V. P., S. Dawre, A. Pandya, L. K. Vora, D. H. Modh, V. Shah et al. 2022. Lyotropic liquid crystals for parenteral drug delivery. J. Control. Rel. 349: 533–549.

Chen, X., W. Wu, L. Liu, J. Hao and S. Dong. 2022. DNA-involved thermotropic liquid crystals from catanionic vesicles. Colloids and Surfaces A: Physicochem. Eng. Aspects 641: 128607.

Chiou, S. and H. J. Shine. 1989. A simplified procedure for preparing 3,5-disubstituted-1,2,4-oxadiazoles by reaction of amidoximes with acyl chlorides in pyridine solution. J. Heterocyclic Chem. 26(1): 125–128.

Choi, E. J., F. Xu, J.-H. Son and W.-C. Zin. 2011. Synthesis and luminescence properties of a lollipop-shaped molecule combined with rod and disc-like mesogens. Liq. Cryst. 551(1): 60–68.

Cristiano, R., F. Ely and H. Gallardo. 2005. Light-emitting bent-shape liquid crystals. Liq. Cryst. 32(1): 15–25.

Das, M. K., B. Barman, B. Das, V. Hamplová and A. Bubnov. 2019. Dielectric properties of chiral ferroelectric liquid crystalline compounds with three aromatic rings connected by ester groups. Crystals 9(9): 473.

Dingemans, T. J. and E. T. Samulski. 2000. Non-linear boomerang-shaped liquid crystals derived from 2,5-bis(p-hydroxyphenyl)-1,3,4-oxadiazole. Liq. Cryst. 27(1): 131–136.

Elston, S. and J. R. Sambles. 1998. The Optics of Thermotropic Liquid Crystals. Taylor and Francis Ltd, London.

Fairhurst, C. E., S. Fuller, J. Gray, M. C. Holmes and G. J. T. Tiddy. 1998. Lyotropic surfactant liquid crystals in handbook of liquid crystals. Demus, D., J. W. Goodby, G. W. Gray, H. W. Spiess, V. Vill (eds.). Volume 3 High Molecular Mass Liquid Crystals page 341 Wiley-VCH.

Fouad, F. S., T. Ness, K. Wang, C. E. Ruth, S. Britton and R. J. Twieg. 2019. Biphenylyl-1,2,4-oxadiazole based liquid crystals—synthesis, mesomorphism, effect of lateral monofluorination. Liq. Cryst. 46(15): 2281–2290.

Frizon, T. E., A. G. Dal-Bó, G. Lopez, M. M. da Silva Paula and L. da Silva. 2014. Synthesis of luminescent liquid crystals derived from gallic acid containing heterocyclic 1,3,4-oxadiazole. Liq. Cryst. 41(8): 1162–1172.

Gallardo, H., R. Cristiano, A. A. Vieira, R. A. Neves Filho, R. M. Srivastava and I. H. Bechtold. 2008. Non-symmetrical luminescent 1, 2, 4-oxadiazole-based liquid crystals. Liq. Cryst. 35(7): 857–853.

Ghosh, T. and M. Lehmann. 2017. Recent advances in heterocycle-based metal-free calamitics. J. Mater. Chem. C 5(47): 12308–12337.

Gieβelmann, F. and P. Zugenmaier. 1995. Mean-field coefficients and the electroclinic effect of a ferroelectric liquid crystal. Physical Rev. 52(2): 1762–1772.

Girotto, E., I. H. Bechtold and H. Gallardo. 2016. New liquid crystals derived from thiophene connected to the 1,2,4-oxadiazole heterocycle. Liq. Cryst. 43(12): 768–1777.

Girotto, E., J. Eccher, A. A. Vieira, I. H. Bechtold and H. Gallardo. 2014. Luminescent columnar liquid crystals based on 1,3,4-oxadiazole. Tetrahedron 70(20): 3355–3360.

Gray, G. W. and J. W. Goodby. 1984. Smectic Liquid Crystals - Textures and Structures, Leonard Hill, Glasgow and London.

Guo, J., R. Hua, Y. Sui and J. Cao. 2014. Synthesis of 3,5-disubstituted 1,2,4-oxadiazoles and their behavior of liquid crystallines. Tetrahedron Lett. 55(9): 1557–1560.

Han, J., J. Y. Wang, F. Y. Zhang, L. R. Zhu, M. L. Pang and J. B. Meng. 2008. Synthesis and mesomorphic behaviour of heterocycle-based liquid crystals containing 1,3,4-oxadiazole/thiadiazole and thiophene units. Liq. Cryst. 35: 1205–1214.

Han, J., M. Zhang, F. Wang and Q. Geng. 2010. Non-symmetric liquid crystal dimers based on 1,3,4-oxadiazole derivatives: synthesis, photoluminescence and liquid crystal behaviour. Liq. Cryst. 37(12): 1471–1478.

Han, M. J., D. Wei, Y. H. Kim, H. Ahn, T. J. Shin, N. A. Clark et al. 2018. Highly oriented liquid crystal semiconductor for organic field-effect transistors. ACS Central Sci. 4(11): 1495–1502.

He, C. F., G. J. Richards, S. M. Kelly, A. E. Contoret and M. O'neill. 2007. Heterocyclic polycatenar liquid crystals. Liq. Cryst. 34(11): 1249–1267.

Kang, S., Y. Saito, N. Watanabe, M. Tokita, Y. Takanishi, H. Takezoe et al. 2006. Low-birefringent, chiral banana phase below calamitic nematic and/or smectic C phases in oxadiazole derivatives. J. Phys. Chem. B 110(11): 5205–5214.

Khan, B. C. and P. K. Mukherjee. 2021. Isotropic to smectic-A phase transition in taper-shaped liquid crystal. J. Mol. Liq. 329: 115539.

Khare, A., R. Uttam, S. Kumar and R. Dhar. 2022. Nanocomposite system of a discotic liquid crystal doped with thiol capped gold nanoparticles. J. Mol. Liq. 366: 120215.

Kotian, S. Y., C. D. Mohan, A. A. Merlo, S. Rangappa, S. C. Nayak and K. M. L. Rai. 2020. Rangappa, K.S. Small molecule based five-membered heterocycles: A view of liquid crystalline properties beyond the biological applications. J. Mol. Liq. 297: 111686.

Kulkarni, S., S. Kumar and P. Thareja. 2021. Colloidal and fumed particles in nematic liquid crystals: Self-assembly, confinement and implications on rheology. J. Mol. Liq. 336: 116241.

Kumar, A. 2022. Dependency of the twist-bend nematic phase formation on the molecular shape of liquid crystal dimers: A view through the lens of DFT. J. Mol. Liq. 354: 118858.

Kuo, H.-M., Y.-L. Chen, G.-H. Lee and C. K. Lai. 2016. Symmetric quinoxaline–oxadiazole conjugates: mesogenic behavior via quinoxaline–CH interactions. Tetrahedron 72(43): 6843–6853.

Lin, K.-T. and C. K. Lai. 2016. Phase crossover in columnar tris-(1,3,4-oxadiazoles) with pendant quinoxalines. Tetrahedron 72(47): 7579–7588.

Lin, K.-T., G.-H. Lee and C. K. Lai. 2015. Mesogenic heterocycles formed by bis-pyrazoles and bis-1,3,4-oxadiazoles. Tetrahedron 71(25): 4352–4361.

Lisa, G., E. R. Cioancă, N. Tudorachi, I. Cârlescu and D. Scutaru. 2011. Thermal degradation of some [1, 3, 4] oxadiazole derivatives with liquid crystalline properties. Thermochim. Acta 524(1-2): 179–185.

Ma, L.-L., C.-Y Li, J.-T. Pan, Y-E. Ji, C. Jiang, R. Zheng et al. 2022. Self-assembled liquid crystal architectures for soft matter photonics. Light Sci. Appl. 11: 270.

Moriya, M., M. Kohri and K. Kishikawa. 2021. Chiral self-sorting and the realization of ferroelectricity in the columnar liquid crystal phase of an optically inactive *N,N'*-diphenylurea derivative possessing six (±)-citronellyl groups. ACS Omega 6(28): 18451–18457.

Nandeesh, K. N., H. A. Swarup, N. C. Sandhya, C. D. Mohan, C. S. Pavan Kumar, M. N. Kumara et al. 2016. Synthesis and antiproliferative efficiency of novel bis(imidazol-1-yl)vinyl-1,2,4-oxadiazoles. New J. Chem. 40(3): 2823–2828.

Nguyen, J., W. Wonderly, T. Tauscher, R. Harkins, F. Vita, G. Portale et al. 2015. The effects of lateral halogen substituents on the low-temperature cybotactic nematic phase in oxadiazole based bent-core liquid crystals. Liq. Cryst. 42(12): 1754–1764.

Niori, T., T. Sekine, J. Watanabe, T. Furukawa and H. Takezoe. 1996. Distinct ferroelectric smectic liquid crystals consisting of banana shaped achiral molecules. J. Mater. Chem. 6: 1231–1233.

Pansu, B. 1999. Are surfaces pertinent for describing some thermotropic liquid crystal phases. Mod. Phy. Lett., B 13: 769–782.

Parra, M., P. Hidalgo, J. Barberá and J. Alderete. 2005. Properties of thermotropic liquid crystals induced by hydrogen bonding between pyridyl-1,2,4-oxadiazole derivatives and benzoic acid, 4-chlorobenzoic acid or 4-methylbenzoic acid. Liq. Cryst. 32(5): 573–577.

Patel, K. D., S. M. Prajapati, S. N. Panchal and H. D. Patel. 2014. Review of synthesis of 1,3,4-oxadiazole derivatives. Synth. Commun. 44(13): 1859–1875.

Percec, V. and D. Sahoo. 2022. Discotic liquid crystals 45 years later. Dendronized discs and crowns increase liquid crystal complexity to columnar from spheres, cubic Frank-Kasper, liquid quasicrystals and memory-effect induced columnar-bundles. Giant 12: 100127.

Prabhu, D. D., N. S. Kumar, A. P. Sivadas, S. Varghese and S. Das. 2012. Trigonal 1, 3, 4-oxadiazole-based blue emitting liquid crystals and gels. J. Phys. Chem. B 116(43): 13071–13080.

Prakash, J., S. Khan, S. Chauhan and A. M. Biradar. 2020. Metal oxide-nanoparticles and liquid crystal composites: A review of recent progress. J. Mol. Liq. 297: 112052.

Qu, S. and M. Li. 2007. Columnar mesophases and phase behaviors of novel polycatenar mesogens containing bi-1,3,4-oxadiazole. Tetrahedron 63(50): 12429–12436.

Rai, K. M. L., N. Liganna, A. Hassner and C. A. Murthy. 1995. Chloramine-T in Organic synthesis. A simple procedure for the synthesis of amino oxadizoles. J. of Science Society of Thailand 22: 71–74.

Reinitzer, F. 1989. Contributions to the knowledge of cholesterol. Liq. Cryst. 5:1, 7–18. https://doi.org/10.1080/02678298908026349.

Saha, S. K., M. K. Paul, A. Chandran, P. K. Khanna and A. M. Biradar. 2017. Low-temperature nematic phase in asymmetrical 1,3,4-oxadiazole bent-core liquid crystals possessing lateral methoxy group. Liq. Cryst. 44(11): 1739–1750.

Salanger, J. L. 1999. In "Surfactants - Types and Uses", Ed.: Laboratorio FIRP, Escuela de Ingenieria Quimica, Universidad de Los Andes, Merida 5101, Venezuela.

Santos, A. B. S., A. M. Manfredi, C. A. M. Salla, G. Farias, E. Girotto, J. Eccher et al. 2021. Highly luminescent liquid crystals by connecting 1,3,4-oxadiazole with thiazolo [5,4-d] thiazole units. J. Mol. Liq. 321: 114887.

Saupe, A. 1973. Disclinations and properties of the director field in nematic and cholesteric liquid crystals. Mol. Cryst. Liq. Cryst. 21: 211–238.

Sebastián, N., M. Čopič and A. Mertelj. 2022. Ferroelectric nematic liquid-crystalline phases. Phys. Rev. E 106: 021C01.

Shanker, G. and C. Tschierske. 2011. Synthesis of non-symmetrically substituted 1,2,4-oxadiazole derived liquid crystals. Tetrahedron 67(45): 8635–8638.

Simić, K. G., P. Rybak, D. Pociecha, L. Cmok, I. Drevenšek-Olenik, T. Tóth-Katona et al. 2022. Introducing the azocinnamic acid scaffold into bent-core liquid crystal design: A structure–property relationship study. J. Mol. Liq. 366: 120182.

Sofia, I. T., T. A. Geivandova, O. Francescangeli and A. Strigazzi. 2003. Banana-shaped 1,2,4-oxadiazole. Pramana- J. of Phys. 61: 239–248.

Takanishi, Y. 2020. Chiral symmetry breaking in liquid crystals: Appearance of ferroelectricity and antiferroelectricity. Symmetry 12(11): 1900.

Tomi, I. H. R., M. Al-Obaidy and A. H. R. Al-Daraji. 2017. Synthesis, characterisation and mesomorphic behaviours of non-symmetrically substituted 1,2,4-and 1,3,4-oxadiazole derivatives. Liq. Cryst. 44: 603–608.

Tschierske, C. 2013. Development of structural complexity by liquid-crystal self-assembly. Angew. Chem. Int. Ed. 52(34): 8828–8878.

Watanabe, J., T. Niori, T. Sekine and H. Takezoe. 1998. Frustrated structure induced on ferroelectric smectic phases in the banana-shaped molecular system. Jpn. J. Appl. Phys. L139.

Yu, Y.-B., W.-L. He, Z.-M. Jiang, Z.-F. Yu, L. Ren, Y. Lu et al. 2019. The effects of azo-oxadiazole-based bent-shaped molecules on the temperature range and the light-responsive performance of blue phase liquid crystal. Liq. Cryst. 46(7): 1024–1034.

Chapter 4

Oxadiazole as Electrochromic Materials

1. Introduction

When the oxidation states of electrochromic materials are changed by applying an electrochemical input, the color change occurs reversibly. The phenomenon of electrochromism was first reported by Deb using WO_3 (Deb 1969; Deb 1973). In recent years, these materials have attracted much attention for various applications, including smart windows, electrochromic displays, and optical switches (Mortimer 1997; Irie et al. 2014; Seeboth et al. 2014). Materials based on conjugated polymers like poly(3-alkylthiophene) and poly(3,4-alkylenedioxythiophene) derivatives (Beaujuge and Reynolds 2010; Akpinar et al. 2013; Tang et al. 2014; Zhang et al. 2016; Brooke et al. 2017), transition metal oxides (Dyer and Leach 1978; Niklasson and Granqvist 2007; Jin et al. 2009), metal complexes (Joseph et al. 1991; Talagaeva et al. 2016; Yan et al. 2016) and organic dyes (Argazzi et al. 2004; Palenzuela et al. 2014) have displayed tremendous potential as electrochemical materials due to intense coloration. However, compared to materials with a colorless neutral state, the sharpness of color transition is substantially lower (Mortimer 1999; Rosseinsky and Mortimer 2001). Recently, electrochromic materials based on poly(3-alkylthiophene) and poly(3,4-alkylenedioxythiopene) derivatives have been reported, which can be employed as flexible electrochromic devices (Mortimer 1999; Rosseinsky and Mortimer 2001) with advantages like low operation voltage, short switching time, good coloration with high modification flexibility in tuning the color (Chang et al. 2008; Oßwald et al. 2017). Recently, a copolymer of oxadiazole and oligo(9,9-dicoctylfuorene) has been reported to show a very sharp color change compared to other electrochromic materials (Ding et al. 2002). Therefore, this chapter is focused on electrochromic materials based on oxadiazole.

2. Mechanism of Electrochromism

Electrochromic materials, in response to an electrical stimulus, undergo reversible changes in color or fluorescence signal due to their oxidation and reduction due to a change in the band gap. The change in band gap causes changes in the optical

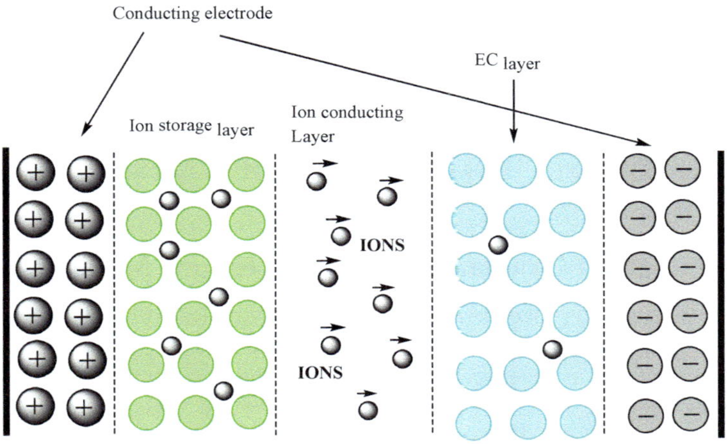

Figure 1. A schematic diagram of an electrochromic device having five layers.

signal (colorimetric or fluorescence). However, continuous electrical input is usually desired to sustain a change in color (or fluorescence). Due to the memory effect, the phenomena require almost negligible to nil current consumption in some electrochromic materials. The electrolyte and electrodes are placed in close contact with the electrochromic material to allow the flow of current in a practical device having a thin film of electrochromic material. The thin film of electrochromic material adopts a multilayer structure that enables the tailoring of properties of the device through the application of optical properties through the application of voltage. Alteration of the polarity of voltage leads to reversal of color change. Practical electrochromic devices usually consist of superimposed layers, such as the transparent conducting oxide layer, ion storage layer, ion conducting layer, electrochromic layer, and the transparent conducting layer (Figure 1). The design has an electrochromic layer coated on one side and an ion storage layer on the other side of the conductor. The ion storage primarily drives charge transport from a power source to the electrochromic material layer (Invernale et al. 2009; Yang et al. 2016; Patel et al. 2017).

The function of an ion conductor, which consists of ionic charge carriers, ensures the circuit completion and facilitates ion transport between electrodes (Marcilla et al. 2006; Thakur et al. 2012; Assis et al. 2015; Moser et al. 2016). The sealant ensures that no electrolyte is leaked when the device functions (Byker 2001; Eh et al. 2018). Transport of H^+ and Li^+ in the devices consisting of a glass of polyester substrate causes an alteration in optical output. Some variations have been observed in laminated devices. For example, the devices dependent on H^+ transport utilize a copolymer of sodium vinylsulfonic acid, 1-vinyl-2-pyrrolidnone, and poly-2-acrylamindo-2-methyl-propane sulfonic acid along with polyethylene oxide (PEO) as electrolyte (Jelle et al. 1993; Ho 1999; Jelle and Hagen 1999; Chen and Ho 2001). Laminated devices are dependent on Li^+ transport, like those consisting of PMMA (polymethyl methacrylate) with polypyrrole (Ericson et al. 2000; Jang et al. 2001; Jang and Oh 2005), polypropylene (Mishra et al. 2005; Nunes et al.

2008), silane (Luo et al. 2008; Puguan et al. 2016), polyvinyl fluoride (Wang et al. 2000; Avellaneda et al. 2002; Barbosa et al. 2010; Jia et al. 2011), propylene carbonate (Bohnke et al. 1982; Zhou et al. 2013; Wen et al. 2014), copolymer of polyethylene glycol methyacrylate with PEO (Munro et al. 1998; Barbosa et al. 2010) and/or glycidyloxypropyl trimethoxysilate with tetraethylene glycol (Munro et al. 1998; Möller et al. 2004) for color modulation function after addition of optimum level of Li salt.

3. Advantages of Oxadiazole as Electrochromic Materials

A variety of materials, like organic polymers, inorganic oxide, or copolymers of hybrid materials, have been used as electrochromic materials (Hedrick and Twieg 1992). Among them as heterocyclic components, the introduction of segments like oxadiazoles impart unique advantages to the materials, which include their high thermal stability, high luminescence quantum yield in the visible region, convenient introduction, processability into materials and robustness of film or thin layers (Hwang and Chen 2002). It is pertinent to mention here that some oxadiazole derivatives are also difficult to process, leading to their decomposition and low solubility in the solvent. The limitations can be overcome by substituting oxadiazole-based materials and their precursors via simple chemical reactions (Liou et al. 2005; Wu et al. 2005; Yen and Liou 2009; Kung and Hsiao 2011). The introduction of bulky triaryl amine substituents usually results in an amorphous polymer with excellent processability and solubility in organic solvents. Oxadiazole derivatives are also employed as efficient electron transporting segments conjugated polymers in one of the bilayer systems (Yen and Liou 2009).

In the following sections, a variety of electrochemical materials with oxadiazole segments have been discussed:

- Oxadiazole in Combination With Polythiophene
- Oxadiazole in Combination With TTF Derivatives
- Oxadiazole in Combination With Carbazoles
- Oxadiazole in Combination With Organic Amines
- Oxadiazole in Combination With Fluorene

3.1 Oxadiazole in Combination With Polythiophene

Conducting polymers like polyacetylene has been the focus of attention due to easy alteration in their band gap, processibility, structure modification, flexibility, and cost-effectiveness (Dyer et al. 2010; Hacioglu et al. 2014). However, its insolubility, instability in air, and high conductivity prevent its application in the electronic industry, which led to search for conjugated polymers with more stability for utility in applications like organic light-emitting diodes (OLED), organic field effect transistors, solar cells, biosensors and electrochromic devices (McQuade et al. 2000; Bolduc et al. 2010; Kamtekar et al. 2010; Li et al. 2010; Mondal et al. 2010; Seo et al. 2011; Ameri et al. 2013). The band gap in conjugated polymers determines their color, conductivity, and optoelectronic properties. These properties can be tubed by planarity, resonance, and bond length alteration along the polymer chain and through interchain donor-acceptor properties (Bouffard and Swager 2008). The past few years have seen the use of benzotriazole, benzothiadiazole, benzoquinolisalines, cayanovinylenes, benzimidazoles, thienopyrazines, perylenes, benzotriazoles, etc., as acceptor unit for turning the band gap in D-A type polymers (Blanchard et al. 2001; Thomas et al. 2004; Aldakov et al. 2005; Durmus et al. 2007; Balan et al. 2008; Gunbas et al. 2008; Balan et al. 2009; Çetin et al. 2009; Karsten et al. 2009; Kozma et al. 2010; Ozelcaglayan et al. 2012; Zaifoglu et al. 2012). Benzooxadiazole is known to be a strong electron acceptor compared to benzotriazole and benzothiadiazole owing to the higher electronegativity of the oxygen atom. Benzooxadiazole utility is further enhanced by coplanarity of its structure, ability to form quinoid structure, air stability, and appropriate band gap. Therefore, it was investigated as an efficient aromatic acceptor moiety in D-A type polymers (Ding et al. 2011; Jiang et al. 2011; Özkut et al. 2012; Jiang et al. 2013). It has also been reported in the literature that the presence of substituents at positions 5 and 6 of benzo oxadiazole provides a polymer with better solubility in common solvents, while the introduction of alkyl chains on the benzo oxadiazole ring yields polymers of high molecular weights and planar conformation (Ding et al. 2011; Jiang et al. 2011; Özkut et al. 2012; Jiang et al. 2013). In a similar context, the presence of a thiophene unit in the material is known to impart a strong and broad absorption band, ensure appropriate energy levels for efficient charge transfer properties, and provide higher stability polymer with high contrast and fast switching time, which makes them appropriate for use in electrochromic devices (Kumar et al. 1998; Balan et al. 2011). Therefore, conjugation of electron-rich donor thiophene in a D-A type structure yields a red-shifted absorption band and enables electrochemical polymerization owing to its low oxidation potential (Sotzing et al. 1996; Irvin and Reynolds 1998). Given this, Özkut et al. have synthesized three D-A polymers based on POP-C10, EOE, and TOT monomers and investigated their electrochromic properties (Özkut et al. 2012). The color produced by polyelectrochomes has been reported to be developed efficiently using the donor-acceptor-donor approach (Havinga et al. 1992).

The three monomers consist of benzo[c][1,2,5]oxadiazole as the acceptor unit and 3,3-didecyl-3,4-dihydro-2H-thieno[3,4-b]]1,4]-dioxepine,3,4-ethylenedioxythiphene or thiophene as the donor unit. The corresponding polymers P(POP-C10), P(EOE), and P(TOT) were investigated for optical and electrochemical output. The electrochemical properties of the monomeric form of EOE and TOT were analyzed in acetonitrile using tetrabutylammonium hexafluorophosphate as the electrolyte, while POP-C10 dichloromethane was used to observe irreversible oxidation peaks at 1.09 (POP-C10), 1.01 (EOE), 1.35 V (TOT) (Table 1). As

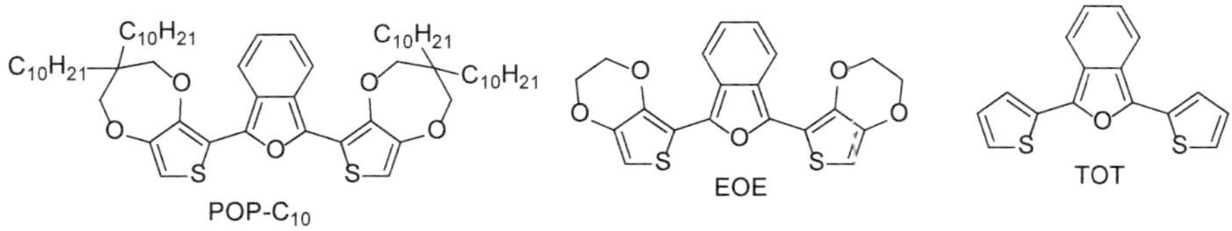

Structures POP-C10, EOE and TOT

Table 1. Electrochemical and optical properties of P(POPC10), P(EOE), and P(TOT), as well as electrochemical properties of their monomers. Reproduced here with the permission of ACS (Özkut et al. 2012).

polymers	$E^{ox}_{m,a}$ (V)	$E^{red}_{m,1/2}$ (V)	$E^{ox}_{p,1/2}$ (V)	$E^{ox}_{p,1/2}$ (V)	$E^{red}_{p,1/2}$ (eV)	E_g^{SPEL} (eV)
P(POP-C$_{10}$)	1.09	−1.3	0.60	−1.3	1.48	1.40
P(EOE)	1.01	−1.2	0.20	−1.24	1.08	1.27
P(TOT)	1.36	−1.1	1.33	−1.06	1.62	1.48

expected for the D-A-D systems, the energy of HOMO level changes with a change in the donor, while not much change in the energy of LUMO is expected in such situations. So, since the acceptor unit (oxadiazole) is the same in each monomer, it is expected that reduction peak potential values shown by the three monomers should be similar, while energies of LUMO levels should alter due to change in donor unit as observed for TOT. It should be noted that cathodic peak potential values are lower than expected. The results observed for EOE, POP-C10, and TOT follow expected trends for such systems (Kaya et al. 2011). The monomers were electrochemically polymerized under conditions similar to those used for cyclic voltammetry to witness a new reversible redox couple, indicating the formation of conducting film on the surface of the electrode. Moreover, an increase in current at the redox couple suggested the formation of electroactive polymers on the surface of the working electrode, which increased in thickness with an increased number of cycles (Figure 2).

The polymer films were studied using cyclic voltammetry. No n-doping properties on adding selenadiazole, thiadiazole, and triazole to P(POP-C10) polymer was witnessed previously, perhaps due to steric effect caused by long alkyl groups in polymer chains (Havinga et al. 1992; İçli et al. 2010). However, polymer P(POP-C10) displayed a peak at –1.5 V during the cathodic scan, indicating the polymer can be p-doped, as shown in Figure 2(d), which may be attributed to the presence of heteroatoms in these monomers. The benzo[c][1,2,5]oxadiazole having electronegative heteroatom was observed to be the most effective acceptor. For polymers and monomers, similar behavior was observed during the oxidation scan due to its dependence on the electron density of the donor moiety [Figure 2(d)]. UV-vis spectral profile for P(POP-C10) revealed absorption bands at 400 nm and 697 nm, P(EOE) revealed absorption bands at 420 nm and 775 nm, while for P(TOT) absorption bands at 345 nm and 560 nm were witnessed. On oxidation, a diminishing effect on the absorbance values of these absorption bands was observed, and the effect was greater on P(EOE) and P(TOT) as compared to P(POP-C10). During oxidation, a new absorption band appeared at 900 nm, P(POP-C10), 1,000 nm, P(EOE), and 750 nm (P(TOT), indicating the generation of charge carriers. For P(POP-C10), P(EOE) and P(TOT) optical band gaps were estimated as 1.40, 1.27, and 1.48 eV, respectively, using UV-vis spectroscopic data, which was observed to agree with the electrochemical data (Table 1). The neutral state of polymer P(EOE) displayed a greenish-blue color, P(POP-C10) displayed cyan, while a magenta color was displayed by P(TOT). After n-doping, P(EOE) displayed a light brown color, P(POP-C10) displayed a brown color, while a dark gray color was displayed by

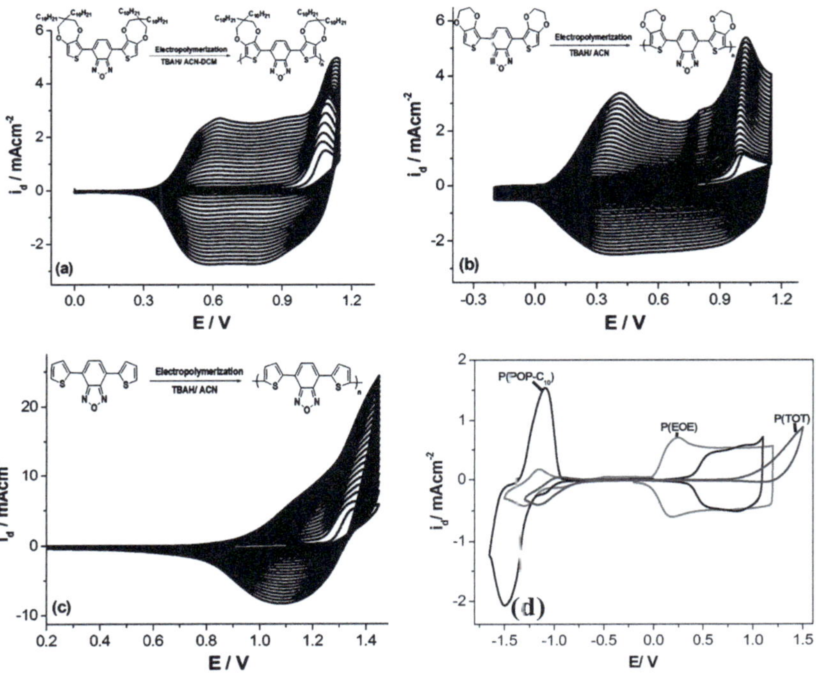

Figure 2. Repetitive cyclic voltammograms (CVs) of (a) 1.5×10^{-3} M POP-C10 in 0.1 M TBAH–DCM/ACN (5:95 v/v), (b) 5.0×10^{-3} M of EOE, and (c) 1.5×10^{-2} M of TOT in 0.1 M TBAH/ACN at a scan rate of 100 mV/s by potential scanning to get P(POP-C10), P(EOE), and P(TOT), respectively. (d) CVs of P(POP-C10), P(EOE), and P(TOT) in 0.1 M TBAH/ACN electrolyte solution at a scan rate of 100 mV/s vs. Ag/AgCl. Reproduced here with the permission of ACS (Özkut et al. 2012).

Table 2. Colorimetric data for P(EOE), P(POP-C10), and P(TOT) polymers on ITO according to CIE (the Commission Internationale de l'Eclairage-International Commission on Illumination) method developed in 1976. Reproduced here with the permission of ACS (Özkut et al. 2012).

Polymers	Colorimetric results				Colors of Polymers		
		Neut.	Ox.	Red.	Neut.	Ox.	Red.
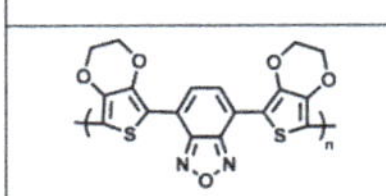 P(EOE)	L*	64.99	76.17	65.27			
	a*	-30.58	1.02	10.24			
	b*	-0.73	-8.76	6.33			
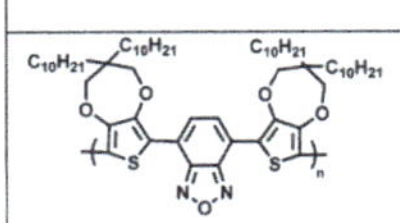 P(POP-C_{10})	L*	59.96	82.12	59.89			
	a*	-19.69	3.93	6.46			
	b*	-16.02	-0.71	7.95			
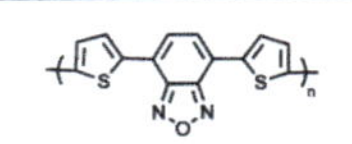 P(TOT)	L*	42.82 [¥]	57.65	44.19			
	a*	51.52 [¥]	-8.76	-0.13			
	b*	0.72 [¥]	-8.09	-2.40			

[¥]These data are measured in the THF solution different from the others polymers' conditions, for this compound data measured on ITO are $L^* = 38.19$, $a^* = 11.63$, and $b^* = -25.95$.

P(TOT) (Table 2). Moreover, a change in color from greenish-blue region to magenta can be accomplished by changing the donor units. Further, using color mixing theory, it was demonstrated that all colors in the visible region of the spectrum may be obtained using these polymers as they exhibit both p- and n-type doping properties, making them excellent candidates for incorporation into electrochromic devices. In addition, the polymer bearing propyledioxythiohene donor units displayed superior properties like high stability, electroactivity even after 2,000 cycles, solubility, high coloration efficiency, and lowest band gap at 1.08 eV.

Given this, Göker et al. have synthesized 3 D-A polymers (PBODT, PBOHT, and PBODHT) containing benzooxadiazole units through electrochemical polymerization and investigated their electrochromic properties (Scheme 1) (Göker et al. 2014). The electropolymerization of the monomers, namely BODT, BOHT, and PBODHT, was performed in CH_2Cl_2/MeCN (5/95, v/v) using 0.1 M sodium perchlorate/lithium perchlorate as supporting electrolyte and observed an enhancement in current density at anode and cathode suggesting deposition of polymer film on working electrode surface (1.25 V/0.93 V for PBODT, 1.23 V/0.88 V for PBOHT and 0.95/–0.49 V for PBODHT; Figure 3). The presence of electron-rich 3,4-ethylenedioxythiophene (EDOT) moiety in PBODHT polymer film reduced redox potential output compared to PBOT and PBOHT polymer. The cyclic voltammetric data was used to obtain energies of HOMO and LUMO energy levels. The HOMO and LUMO energy levels

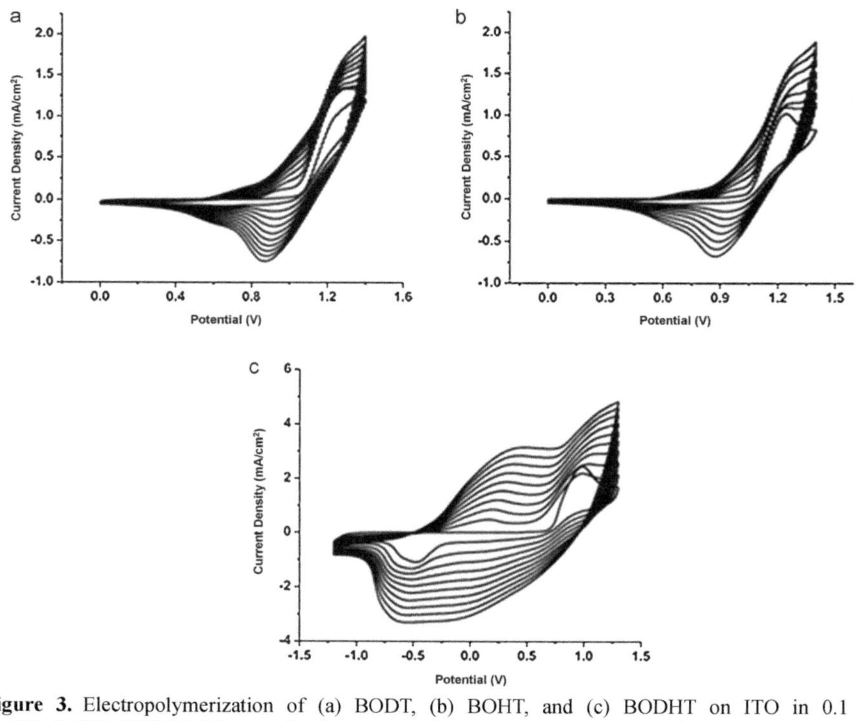

Figure 3. Electropolymerization of (a) BODT, (b) BOHT, and (c) BODHT on ITO in 0.1 M $NaClO_4/LiClO_4$/DCM/ACN electrolyte/solvent couple at a scan rate of 100 mV s^{-1}. Reproduced here with the permission of Elsevier (Göker et al. 2014).

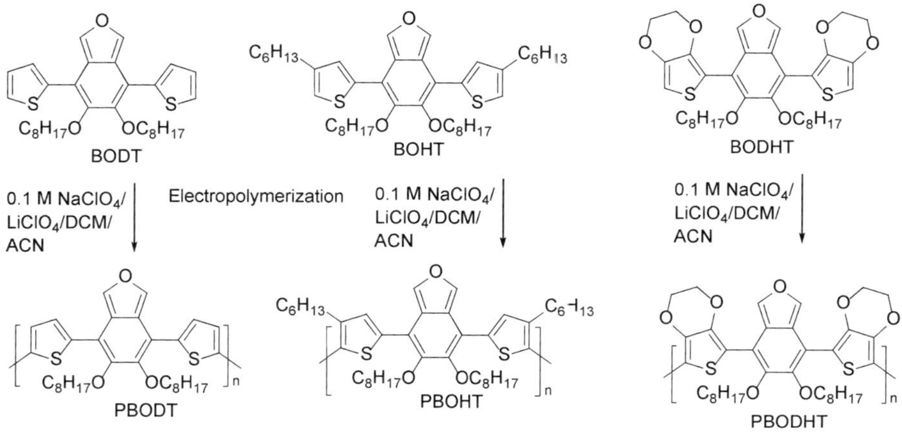

Scheme 1. Synthesis of polymers PBOTDT, PBOHT, and PBODHT.

were estimated at –5.73 eV and –6.98 eV for PBODT. In the case of PBOHT, 5.54 eV energy was estimated for the HOMO orbital, while –5.54 eV energy was estimated for the LUMO orbital. Similarly, –4.43 eV and –6.23 eV energies were estimated for HOMO and LUMO energy levels of PBODHT, respectively.

Spectro-electrochemistry experiments on examination revealed optical and electronic changes like band gap, intergap states, absorption maxima, and formation of polaronic and bipolarnic bands upon doping. The formation of polarons and bipolarons was investigated by acquiring UV-vis-NIR spectra as a function of applied potential. The formation of lower energy carriers occurs due to absorbance enhancement in the UV-vis absorption spectra of the polymer films. The potential was scanned in 0.0 to 1.2 V region for PBODT polymer, 0.0 V to 1.15 V for PBOHT, and –1.20 V to 0.8 V for PBODHT during polymerization on ITO glass slide, which eliminates the possibility of dopant ions and charges formed during the electropolymerization of monomers. The polymers were initially reduced to neutral states, followed by oxidation during spectroelectrochemistry studies. The studies revealed absorption maxima for PBODT at 551 nm, for PBOHT at 577 nm, and for PBODHT at 590 nm, which were used to estimate band gaps as 1.25, 1.27, and 1.50 eV for polymers PBODT, PBOHT, and PBODHT, respectively (Figure 4). The difference in band gap was attributed to steric interactions of alkyl chains at the third position of the thiophene unit, which distorted the planarity of the polymer main chain in these polymers, resulting in reduced overlap between π-orbitals. However, bipolarons and polarons bands were not resolved, perhaps due to bipolarons not forming through oxidation of polarons but due to polarons coupling.

Due to the utility of electrochromism in many commercial applications like displays, windows, mirrors, and sunglasses, p-doping/dedoping studies were performed. The polymer possessing thiophene switched between purple and blue colors, the polymer with 3-hexylthiophene switched between blue and pale green colors, and the polymer possessing EDOT switched between dark blue and pale blue. In the scientific representation of color as per CIE (Commission of internationale de

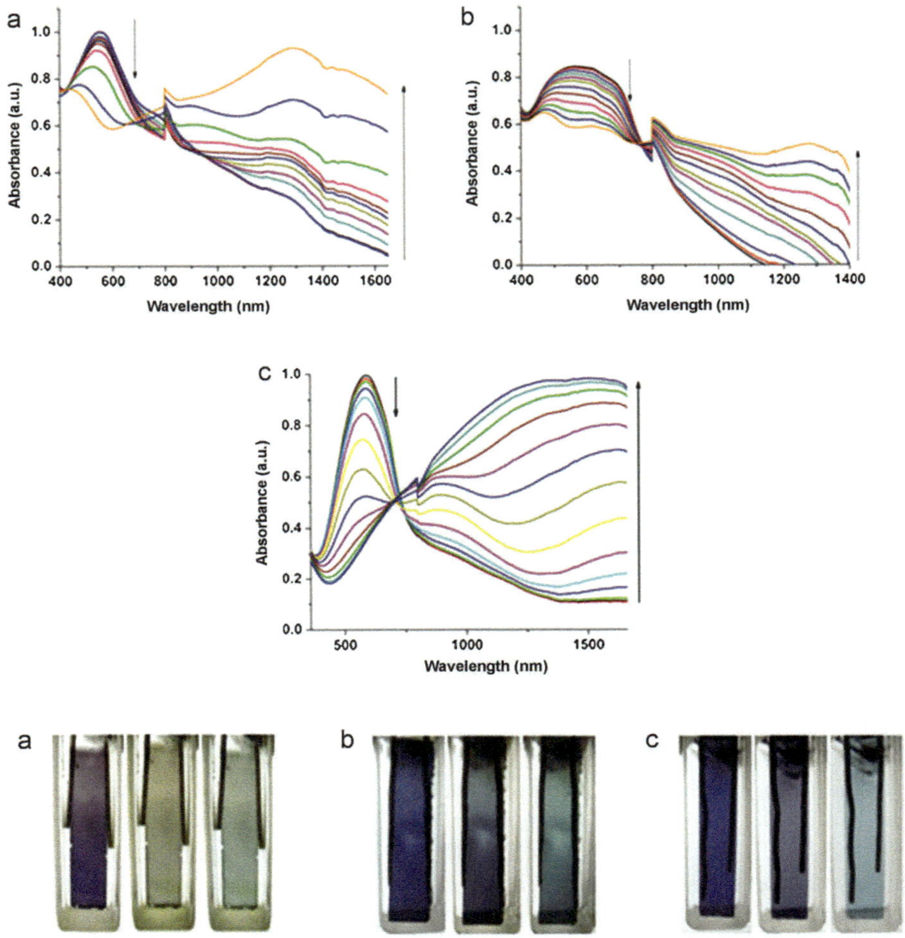

Figure 4. Change in the electronic absorption spectra and colors of (a) PBODT, (b) PBOHT, and (c) PBODHT upon oxidative doping at potentials between 0.0/1.40 V, 0.0/1.25 V and –1.20/+1.0 V, respectively. Reproduced here with the permission of Elsevier (Göker et al. 2014).

l' Eclairage), *L* denotes the brightness of the color (L = 0 provides black and L = 100 denotes diffuse white), *a* denotes a color between red/magenta, and *b* denotes a color between yellow/blue. Therefore, measurements were performed to determine these parameters for the polymers (Table 3).

The optical values of PBODHT were witnessed between PBODT and PBOHT with 75% transmittance change at 590 nm, while 88% transmittance change at 1,500 nm was observed in the visible region. Moreover, EDOT and thiophene-containing polymers display a rapid switching between their neutral and oxidized states. The polymer PBODT displays switching times of 1.1 s at 555 nm and 0.5 s at 1,300 nm. The polymer PBOHT displayed a switching time of 2.5 s at 580 nm and 1.5 s at 1,290 nm, while PBOHT displayed a switching time of 0.9 s at 590 nm and 1.1 s at 1,500 nm. Therefore, it can be seen from the data that the polymer PBODHT displays

Table 3. Colorimetric measurements of the polymer films. Reproduced here with the permission of Elsevier (Göker et al. 2014).

PBODT	0.0 V	1.2 V	1.4 V
	L: 55.786	L: 76.860	L: 75.451
	a: 10.048	a: −9.822	a: −16.152
	b: −13.948	b: 19.619	b: 9.624
PBOHT	0.0 V	1.2 V	1.4 V
	L: 42.308	L: 51.496	L: 57.751
	a: 3.222	a: −2.489	a: −17.344
	b: −26.159	b: −6.410	b: 8.169
PBODHT	−1.2 V	0.5 V	0.7 V
	L: 31.190	L: 53.190	L: 61.190
	a: 3.452	a: −2.890	a: −21.760
	b: −24.210	b: −7.978	b: −5.132

the fastest switching times and highest contrast among the three polymers used in the study. The study suggested that three polymers are p-dopable and display high contrast in the NIR region. Moreover, using the EDOT unit as an electron-rich EDOT unit as a donor unit with oxadiazole as an acceptor unit exerts a stronger influence on the electronic properties of the polymers, leading to a red shift in the absorption and a decrease in oxidation potential.

3.2 Oxadiazole in Combination With TTF Derivatives

Tetrathiafulvalene (TTF) derivatives are known for their reversible electrochemical oxidation and exceptional thermodynamic stability of radical cations and dicationic form stability (Bryce 1991). Moreover, their low oxidation potential (TTF, $E^{1}{}_{,1/2}$ = 0.32 V vs. Ag/AgCl) imparts them with properties like low power consumption and low driving voltage, making them attractive for their use in electrochromic devices. Given this discussion, Wang et al. have synthesized three TTF derivatives conjugated to 1,3.4-oxadiazole units were synthesized (TTFOX1-TTFOX3) and demonstrated the potential of such materials as electrochromic devices (Wang et al. 2004). Cyclic voltammograms of the oxadiazole-TTF combinations (TTFOX1-TTFOX3) were acquired in DCM using 0.1 M tetrabutylammonium hexafluorophosphate (TBAPF6) as a supporting electrolyte. The TTFOX1-TTFOX3 combinations displayed reversible TTF type two-step, one-electron redox couples in 0 to 1 V region (vs. Ag/AgCl). A comparison of the half-wave potentials of TTFOX1 ($E^{1}{}_{,1/2}$ = 0.12 and $E^{2}{}_{,1/2}$ = 0.63 V) and TTFOX2 ($E^{1}{}_{,1/2}$ = 0.16 and $E^{2}{}_{,1/2}$ = 0.64 V) with TTF1 (Bryce 2000; Segura and Martín 2001) suggested a 100 mV anodic shift in the half-wave potential owing to electron acceptor properties of oxadiazole derivative. Moreover, the presence of ethynylene segment in TTFOX2 led to a small variation in electrochemical signal as compared to TTFOX1. The presence of an additional oxadiazole unit and absence of methyl groups (electron donating) led to the highest shift in anodic potential in TTFOX3 ($E^{1}{}_{,1,2}$ = 0.31 V and $E^{2}{}_{,1/2}$ = 0.74 V) as compared to TTF1. The electrochromic properties of TTFOX1-TTFOX3 were investigated using spectroelectrochemistry, and absorption maxima of the oxidized and neutral forms are listed in Table 4.

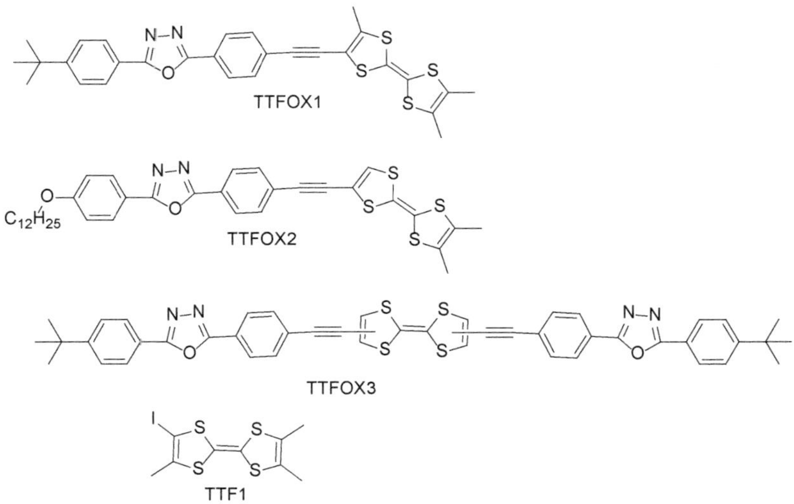

Structures of TTFOX1, TTFOX2, TTFOX3 and TTF1

Table 4. UV-visible absorption maxima (nm) of TTFOX1–TTFOX3 and TTF in neutral and their charged states. Referenced (Hünig et al. 1973; Ashton et al. 1999). Referenced (Ashton et al. 1999). Reproduced here with the permission of RSC (Wang et al. 2004).

TTFOX1	447		
TTFOX1$^{+\cdot}$	437, 468		703
TTFOX1^{2+}		520	
TTFOX2	462		
TTFOX2$^{+\cdot}$	468	596	669, 714
TTFOX2^{2+}		507	
TTFOX3	447		
TTFOX3$^{+\cdot}$	435		722
TTFOX3^{2+}		585	
TTF^{b}	450		
TTF$^{+\cdot b}$	438	580	
TTF^{2+c}	*ca.* 350^{d}		

[c] DCM solution containing 0.1 M $TBAPF_6$.
[c] Under our conditions, **TTF**$^{2+}$ precipitated and the spectrum could not be determined clearly.

The UV-vis spectra of TTFOX3 in DCM using 0.1 M TBAPF6 were recorded as a function of applied voltage (vs. Pt electrode) (Figure 5). The neutral form of TTFOX3 displayed an absorption band at 447 nm, while the radical cationic form TTFOX3^{+} formed through oxidation displayed two new absorption bands at 435 nm and 722 nm. On increasing the voltage, the two absorption bands gradually decreased in intensity, and a new absorption band at 585 nm appeared, indicating the formation of dicationic form TTFOX3^{2+}. A comparison of absorption bands in TTFOX3 and TTF (Wudl et al. 1970; Hünig et al. 1973; Bryce 1991) disclosed a large red shift in the absorption maxima of the dicationic form (Δ 230 nm) as compared to the radical cationic form (Δ 140 nm), indicating extended conjugation in dicationic form of TTFOX3. Similar optoelectrochemical properties were observed in three TTFOX1-TTFOX3 derivatives. For instance, the neutral form of TTFOX1 displayed

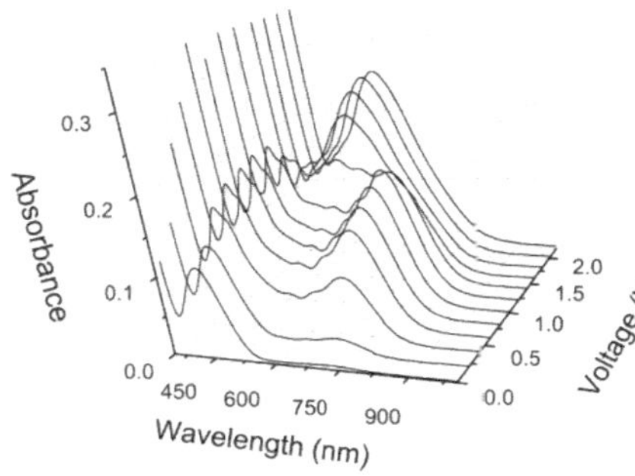

Figure 5. Absorption spectroelectrochemistry of TTFOX3 in DCM having 0.1 M TBAPF6 as the supporting electrolyte. A 1 mm quartz cell equipped with Pt mesh as the anode and Pt wire was used both the counter and reference electrodes was employed. Reproduced here with the permission of RSC (Wang et al. 2004).

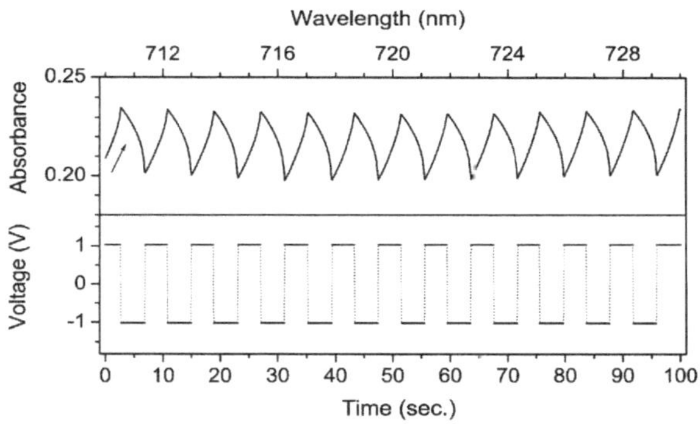

Figure 6. Demonstration of the optical switching in DCM solution of TTFOX3, employing a spectroelectrochemical cell as described in Figure 5. While the spectrometer was scanning across the absorption maxima of $TTFOX3^{+\cdot}$ from 710 to 730 nm, electrical square waves were employed. Reproduced here with the permission of RSC (Wang et al. 2004).

an absorption band at 447 nm, while the high energy band of its radical, cationic form displayed two absorption maxima at 437 nm and 468 nm. The lower energy radical, cationic form displayed an absorption maximum at 703 nm, while a dicationic form of TTFOX1 displayed an absorption band at 520 nm. Similarly, the absorption band of the neutral form of TTFOX2 appeared at 462 nm due to extended p-conjugation of the oxadiazole chromophore. Switching experiments were reported to explore the utility of TTFOX1-TTFOX3 in electrochromic devices (Figure 6). On application of electrical square waves of ±1.03 V (Pt electrode as reference and grounded to pulse generator chassis) across the spectroelectrochemical cell containing a solution of TTFOX3, the optical absorption around the $TTFOX3^{+}$, yielding an altered absorption maximum displaying a reversible color change (dark green/orange) (Figure 6).

3.3 Oxadiazole in Combination With Carbazole Derivatives

Derivatives of carbazoles or polymers consisting of carbazole units have received increasing attention owing to their unique optical properties and ability to transport holes in optoelectronic devices (Rani and Santhanam 1998; Tao et al. 1998). Polymers that consist of hole-transporting (acceptor) units are of great interest due to the possibility of increasing both hole and electron transport affinities in such materials. The properties of materials possessing a carbazole moiety can be easily fine-tuned through functionalization at nitrogen, 3,6, and 2,7 positions. Incorporating oxadiazole units as an electron-deficient acceptor unit into a polymer chain stabilizes the electron-charged state of electrochromic materials. Incorporating oxadiazole units in polymeric electrochromic materials also thermally stabilizes them and prolongs their application in electrochromic devices.

Given the above discussion, Udum et al. have reported a 2,5-bis(9-methyl-9H-carbazol-3-yl)-1,3,4-oxadiazole (CZOX), which was electrochemically oxidized to poly(2,5-Bis(9-methyl-9H-carbazol-3-yl)-1,3,4-oxadiazole) yielding stable films with electrochromic properties (Udum et al. 2015). Spectroelectrochemical analysis was performed, which demonstrated the feasibility of both the p-type and n-type doping processes. The n-dopable properties were demonstrated by polymer through the presence of redox response at negative potential. The thin films of P(CZOX) were potentiodynamically deposited on ITO using a solution of 0.01 M CZOX and tetrabutylammonium fluoroborate in acetonitrile (Scheme 2). Incrementally raising the potential between 0 to 1.5 V led to no sharp absorption peak in the visible range at

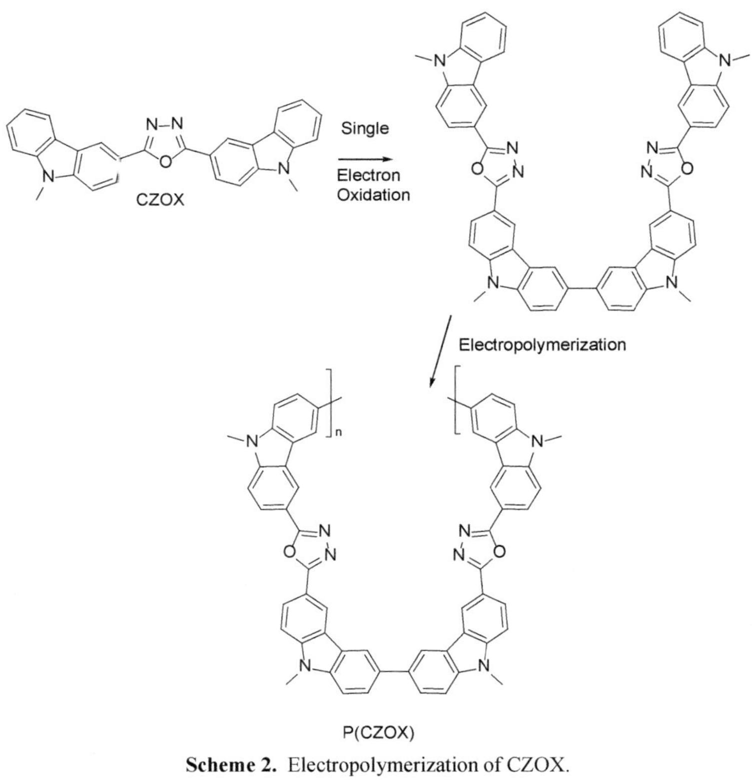

Scheme 2. Electropolymerization of CZOX.

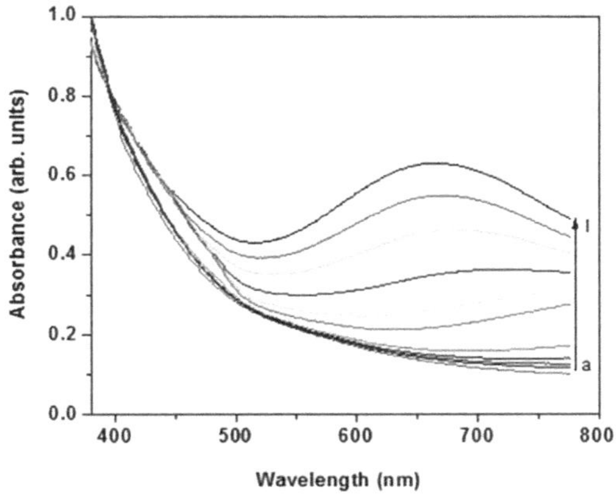

Figure 7. The p-Doping spectroelectrochemistry of poly(CXOZ) film at 660 nm in 0.1 M TBABF4/MeCN on an ITO-coated glass slide in monomer-free solution at applied potentials (V); (a) 0.0 (b) 0.2 (c) 0.4 (d) 0.6 (e) 0.7 (f) 0.8 (g) 0.9 (h) 1.0 (i) 1.1 (j) 1.2 (k) 1.3 (l) 1.4. Reproduced here with the permission of Elsevier (Udum et al. 2015).

the neutral state in Figure 7. Further increase in potential value led to the appearance of a new absorption band near 660 nm, imparting a blue color to the polymer film.

The electrochromic device constructed based on P(CZOX) operated between transmissive gray and blue colored states on the application of a potential between −1.6 V and +1.6 V. The kinetic studies performed with polymer demonstrated a high contrast ratio (75%), excellent stability, color persistence, and fast response time.

Wu et al. have reported carbazole derivatives (COP) having (trifluoromethyl) pyridine and COI having isoxazole conjugated to oxadiazole unit to investigate the electrochromic properties of their oxidized polymeric films (Wu et al. 2023). Cyclic voltammetric studies of both COP and COI (2 mM) in 0.1 M $LiClO_4$ solution indicated E_{onset} as 1.38 and 1.25 V (Vs. Ag/AgCl), respectively. The isoxazole ring having an imine as an electron-withdrawing group represents a weak electron-donating unit, while the 3-(trifluoromethyl)pyridine unit is an electron-withdrawing unit in COP. Therefore, the E_{onset} of COP was observed to be higher than COI. Moreover, a continuation of the CV sweep led to an enhancement in the intensity of redox peaks and the development of electrochemically active P(COP) and P(COI) films on the surface of the ITO working electrode (Scheme 3). The thermal properties of electrosynthesized films of P(COP) and P(COI) were studied using thermogravimetric (TGA) analysis under a nitrogen atmosphere to observe a 5% decomposition at 249°C and 239°C and a char yield of 28% and 44% at 800°C, respectively.

The UV-vis-NIR spectra of P(COP) and P(COI) electrodes in solution are shown in Figure 8(a, b). At a low value of applied potential, no absorption peak is visible in the visible region (> 400 nm). On application of an applied potential of 0.8 V, the oxidation of P(COP) started, leading to the appearance of polarons and bipolarons absorption bands at 423 nm, 710 nm, and 1,100 nm (Kuo et al. 2022). Similar to their corresponding polymers, the monomers COP and COI did not display an absorption band beyond 400 nm. The P(COP) film displayed yellow ocher, grayish green, and

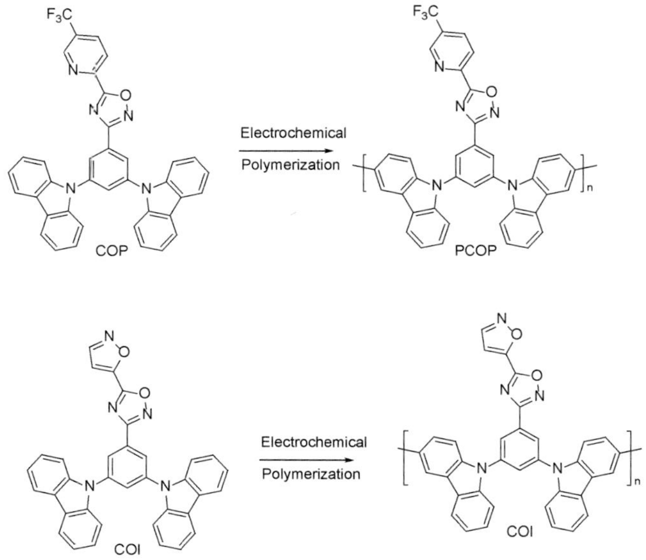

Scheme 3. Electro-polymerization of COP and COI.

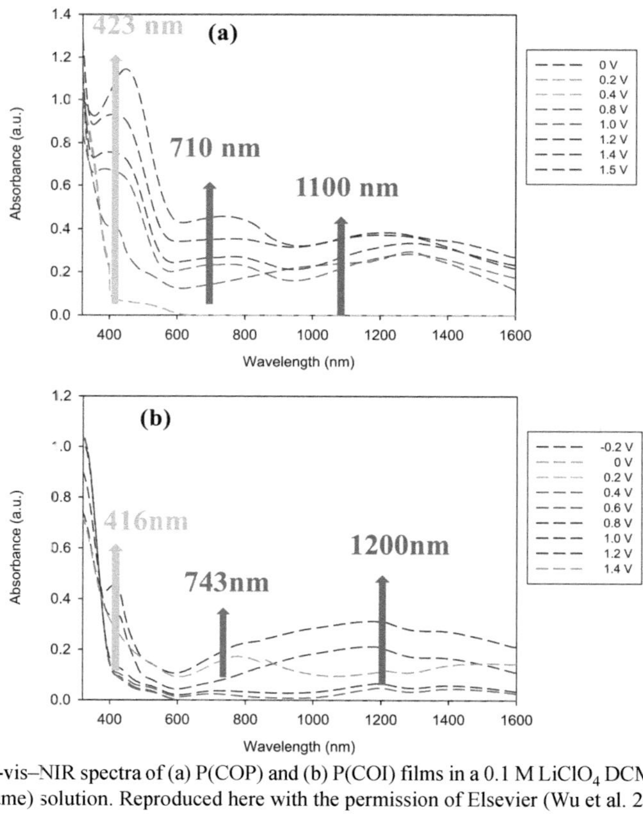

Figure 8. UV-vis–NIR spectra of (a) P(COP) and (b) P(COI) films in a 0.1 M $LiClO_4$ DCM/ACN (1:1, by volume) solution. Reproduced here with the permission of Elsevier (Wu et al. 2023).

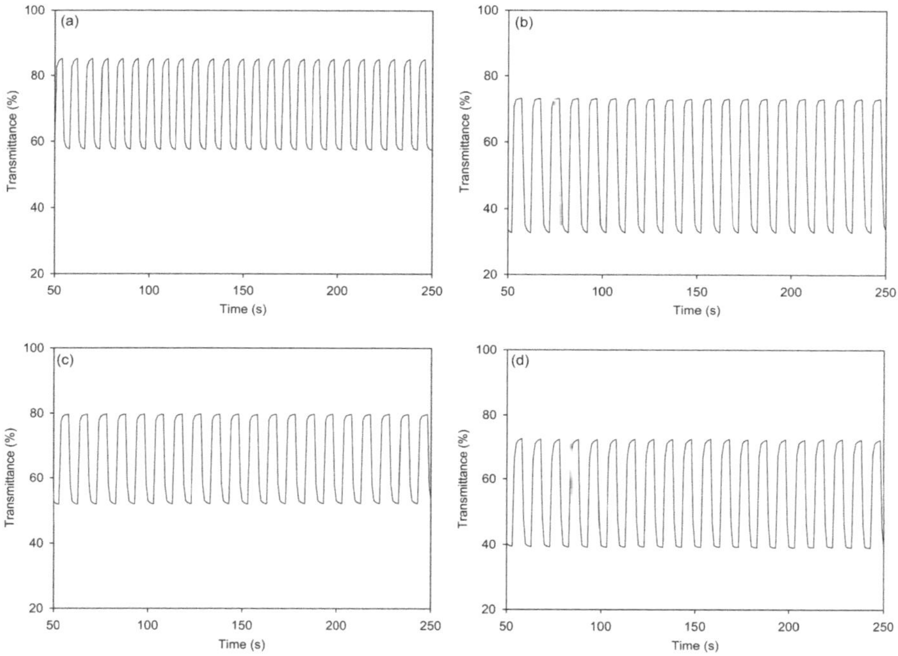

Figure 9. Transmittance changes of (a) P(COP) at 710 nm, (b) P(COP) at 1,100 nm, (c) P(COI) at 743 nm, and (d) P(COI) at 1,200 nm between the colored and bleached state with a time interval of 5 s in a 0.1 M $LiClO_4$ DCM/ACN (1:1, by volume) solution. Reproduced here with the permission of Elsevier (Wu et al. 2023).

khaki green colors at 0.0, 1.4, and 1.5 V potential values, respectively. The film of P(COI) displayed light grayish-green, light-greenish yellow, and light green colors at 0.2, 1.2, and 1.4 V potential values, respectively. The color displayed by films of polymers changes further upon a change in the doped state. The electrochromic switching kinetics P(COP) and P(COI) between potential values of 0.0 and 1.4 V in solution at an interval of 5 s is shown in Figure 9. The change in transmittance value for P(COP) was 27.5% at 710 nm and 40.1% at 1,100 nm, respectively, while for P(COI), the transmittance value at 28.2% at 743 nm and 34.0% at 1,200 nm, respectively were observed. The polycarbazoles containing oxadiazole-based ECDs demonstrated high electrochemical redox stability, short bleaching, and coloring time in the study. The authors concluded that polycabazoles, in combination with oxadiazole, are promising electrochemical materials for use in multicolored and energy-saving electrochromic devices.

3.4 Oxadiazole in Combination With Organic Amines

Polymers having triarylamine are widely used as electrochromic and photoluminescent materials (Lin et al. 2007; Liou et al. 2007). The bulkiness of propeller-shaped triphenylamine units imparts polymer with high thermal stability, good solubility in many organic solvents, and good film-forming ability. Due to the expected

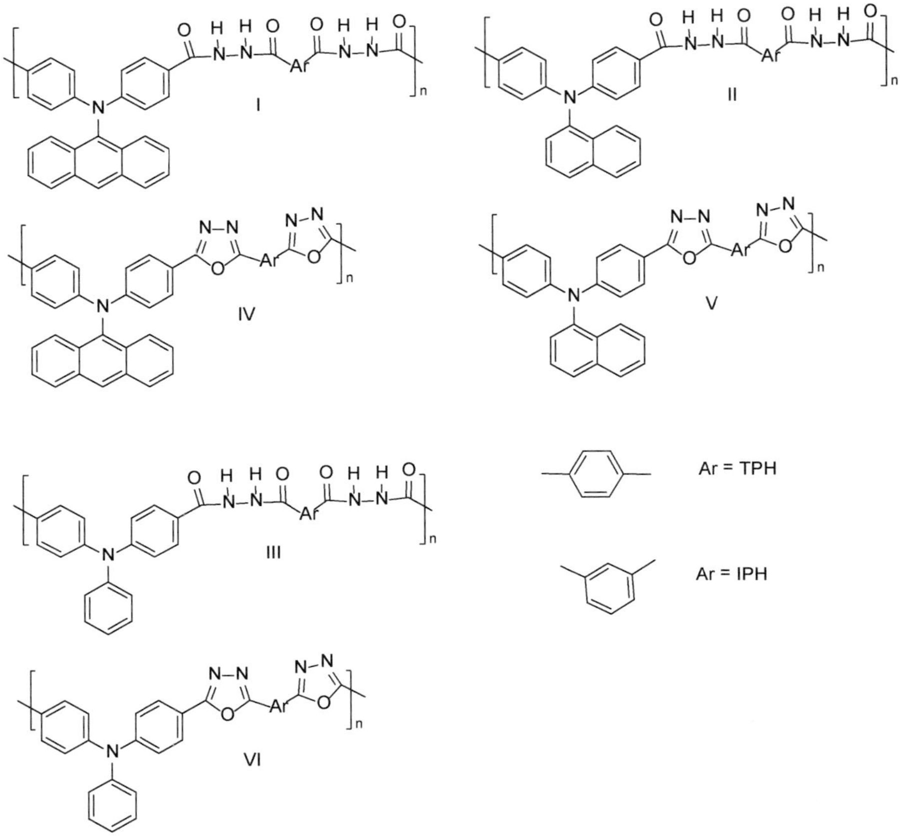

Structures I to VI with IPH and TPH

advantages associated with its coupling to oxadiazole segments, Yen and Liou have synthesized poly(amine-hydrazide) and poly(amine-1,3,4-oxadiazole) derivatives (Yen and Liou 2009). In the study, polymers 1–VI were prepared. The polymers having hydrazides were thermally converted into poly(amine-1,3,4-oxadiazole)s with IPH and TPH spacer units.

The glass transition temperatures of (Tg) of poly(amino-hydrazide)s and poly(amine-1,3,4-oxadiazole)s were reported using differential thermal calorimetry (DSC) to be in the range 200–221°C and 297–318°C, respectively. The thermogravimetric analysis (TGA) suggested no significant weight loss for all poly(amine-1,3,4-oxadiazole)s, indicating their stability. A significant char yield for polymer at around 55% was observed for high aromatic content. The optical properties of poly(amine-hydrazide)s and poly(amine-1,3,4-oxadiazole) were reported using UV-vis and photoluminescence spectroscopy.

They displayed good solubility and absorption bands in the 344–391 nm region in N-methyl-2-pyrrolidone (NMP), which may be assigned to π to π^* transition due to the aromatic chromophore. The photoluminescence spectra displayed bands in the

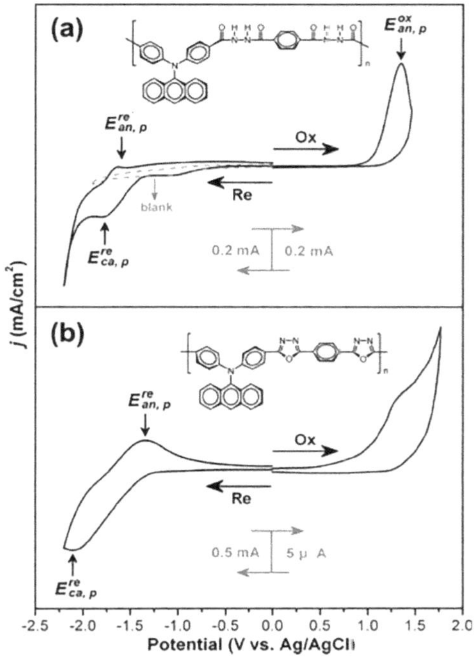

Figure 10. Cyclic Voltammograms of poly(aminehydrazide) (a) I-TPH and poly(amine-1,3,4-oxadiazole) (b) IV-TPH films onto an indium-tin oxide (ITO)-coated glass substrate in CH_3CN (oxidation) and DMF (reduction) solution containing 0.1 M TBAP at a scan rate of 50 and 100 mV/s, respectively. Reproduced here with the permission of (Yen and Liou 2009).

490–497 nm region. The electrochemical properties of poly(amine-hydrazide)s and poly(amine-1,3,4-oxadiazole)s obtained using cyclic voltammetry and differential pulse polarography are listed in Table 5. The cyclic voltammogram of I-TPH and poly(amine-1,3,4-oxadiazole) IV-TPH is shown in Figure 10, while Figure 11 depicts the differential pulse voltammograms of the films of anthrylamine-based polymer. On comparison of the peak potential of IV-TPH with I-TPH, it can be concluded that the second reduction in IV-TPH occurs in the anthracene group, while the third reduction occurs in the oxadiazole moieties. Moreover, the first and second reduction stages in IV-IPH can also be assigned to the anthracene and oxadiazole segments, respectively. The energies of HOMO and LUMO orbitals calculated from the onset of oxidation and reduction potentials along with the onset of absorption wavelength are listed in Table 5.

The study concluded that introducing a bulky anthracene group into the polymer chain increases interchain interactions and disrupts coplanarity, thereby enhancing the solubility of the polymer. The resulting amorphous polymer displayed high optical transparency during UV-vis transmittance measurements in a cutoff wavelength of 415–435 nm region. The poly(amine-hydrazide) I-IPH and poly(amine-1,3,4-oxadiazole) IV-IPH exhibited green fluorescence with a maximum near 490 nm in NMP with high quantum yield (29.9% and 13.5%). The anthracene-based poly(amine-1,3,4-oxadiazole)s displayed electrochromic

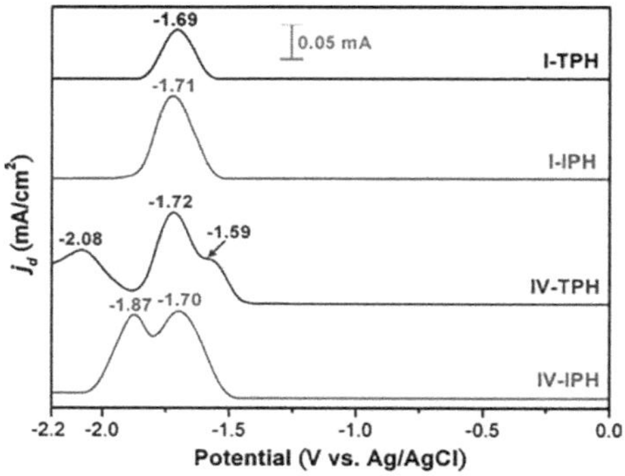

Figure 11. Differential pulse voltammograms of anthryl amine-based polymer films onto an indium-tin-oxide (ITO)-coated glass substrate in DMF solution containing 0.1 M TBAP. Scan rate, 5 mV/s; pulse amplitude, 50 mV; pulse width, 50 ms; pulse period, 0.2 s. Reproduced here with the permission of (Yen and Liou 2009).

Table 5. Electrochemical Properties of different molecule I to VI. Reproduced here with the permission of (Yen and Liou 2009).

	Oxidation/V[a]	Reduction/V[b]					
Index	E_{onset}	E_{onset}	E_g^{EC} (eV)	E_g^{Opt} (eV)	E_{HOMO} (eV)[c]	E_{LUMO}^{Opt} (eV)	E_{LUMO}^{EC} (eV)
I-TPH	1.07	−1.46	2.53	2.84	5.43	2.59	2.90
I-IPH	1.07	−1.46	2.53	2.95	5.43	2.48	2.90
II-TPH	1.11	–[d]	–	3.08	5.47	2.39	–
II-IPH	1.11	–	–	3.11	5.47	2.36	–
III-TPH	1.08	–	–	2.97	5.44	2.47	–
III-IPH	1.08	–	–	2.99	5.44	2.45	–
IV-TPH	1.10	−1.40	2.50	2.76	5.46	2.70	2.96
IV-IPH	1.12	−1.41	2.53	2.88	5.48	2.60	2.95
V-TPH	1.12	−1.49	2.61	2.83	5.48	2.65	2.87
V-IPH	1.12	−1.60	2.72	2.93	5.48	2.55	2.76
VI-TPH	1.09	−1.41	2.50	2.75	5.45	2.70	2.95
VI-IPH	1.09	−1.55	2.64	2.86	5.45	2.59	2.71

E_g^{EC} (Electrochemical band gap): Difference between E_{HOMO} and E_{LUMO}^{EC}.
E_g^{Opt} (Optical band gap): Calculated from polymer films ($E_g = 1240/\lambda_{onset}$).
E_{LUMO}^{Opt} (LUMO energy levels calculated from optical method): Difference between E_{HOMO} and E_g^{Opt}.
[a] vs. Ag/AgCl in CH_3CN.
[b] vs. Ag/AgCl in DMF.
[c] The HOMO and LUMO energy levels were calculated from cyclic voltammetry and were referenced to ferrocene (4.8 eV).
[d] No discernible signal was observed.

characteristics showing a color change from pale yellow (neutral form) to red (reduced form) on change the potential negatively from 0 to –2.20 V. Therefore, it was demonstrated that anthrylamine-based polymers have great potential as electrochromic material with high thermal stability, solubility, quantum efficiency and electrochemical behavior.

3.5 Oxadiazole in Combination With Fluorene

The emission of blue light is normally associated with high energy gaps and requires the application of high electric field intensities to the layers emitting light. Therefore, there is a demand for blue light-emitting polymers with high oxidative and thermal stabilities. In this context, fluorene-based polymers represent interesting blue light-emitting materials that have high chemical stability and allow easy structural tuning to achieve the desired results. A variety of polymers based on fluorene copolymerized with other monomers have been reported with good photoluminescence efficiency and high thermal and oxidative stability. Moreover, functionalization at the 9-position also offers control of interchain interactions, cross-linking ability, and electrical and optical properties. A highly efficient electrochromic material may be obtained using a balance of holes and electrons in the polymer emissive layer. Given this, Ding et al. have reported an alternative copolymer of fluorene and oxadiazole and reported its electrochromic properties (Ding et al. 2010). The alerting polymer PF4OX on electrochemical reduction produced electrochromic properties by acting as an n-type polymer. The presence of oxadiazole segments in the poly(9,9-dioxtylfluorence) chain imparted stability to the polymer and electrochemical reduction, which led to color change from colorless (neutral form) to rosy red at the electron-charged state and displayed a switching time of 1.1 s. The absorption profile of the polymer revealed two-absorption bands at 510 nm and 560 nm. The absorption band at 510 nm was attributed to the polaron on the fluorene unit, while the 560 nm peak was attributed to the polaron on the oxadiazole segment. The polymer film displayed high stability during the electrochromic switching and a 30% loss of transmittance change in 500 potential square wave cycles. FT-IR-based studies revealed the decay of the electrochromic performance of the film on the formation of fluorenone and quinone defects on the fluorene unit in the copolymer chain. Laser flash photolysis-based studies revealed high electron localization potential and vulnerability to attack by oxygen to form the defects. These defects act as charge carrier traps and producte a pair of inner peaks at −1.84 and 1.03 V in the cyclic voltammetry test of copolymer film. The study demonstrated the potential of the polymer for developing an electrochromic device.

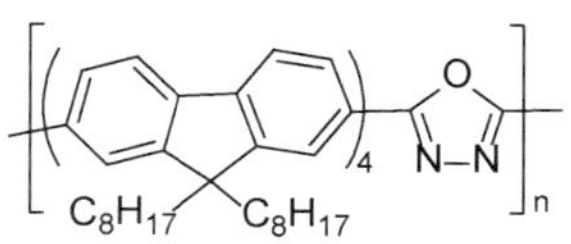

PF4Ox

Structure PF4Ox

4. Conclusion

A variety of electrochromic systems that have oxadiazole as an electron acceptor moiety have been reviewed. Various design parameters have been discussed in the chapter. The use of oxadiazole units with electron donors in D-A configurations has been explored. The p-dopable and n-dopable properties of electrochromic systems have been discussed. Various techniques used for the analysis of such devices have also been discussed. The authors hope this chapter allows readers to review and design improved electrochromic devices based on previously published data presented here.

References

Akpinar, H. Z., Y. A. Udum and L. Toppare. 2013. Spray-processable thiazolothiazole-based copolymers with altered donor groups and their electrochromic properties. J. Polym. Sci., Part A: Polym. Chem. 51(18): 3901–3906.

Aldakov, D., M. A. Palacios and P. Anzenbacher. 2005. Benzothiadiazoles and dipyrrolyl quinoxalines with extended conjugated chromophores–fluorophores and anion sensors. Chem. Mater. 17(21): 5238–5241.

Ameri, T., N. Li and C. J. Brabec. 2013. Highly efficient organic tandem solar cells: a follow up review. Energy Environ. Sci. 6(8): 2390–2413.

Argazzi, R., N. Y. Murakami Iha, H. Zabri, F. Odobel and C. A. Bignozzi. 2004. Design of molecular dyes for application in photoelectrochemical and electrochromic devices based on nanocrystalline metal oxide semiconductors. Coord. Chem. Rev. 248(13): 1299–1316.

Assis, L. M. N. d., J. R. d. Andrade, L. Santos, A. d. J. Motheo, B. Hajduk, M. Łapkowski et al. 2015. Spectroscopic and microscopic study of Prussian blue film for electrochromic device application. Electrochim. Acta 175: 176–183.

Avellaneda, C., K. Dahmouche and L. Bulhoes. 2002. All sol-gel electrochromic smart windows: CeO 2-TiO_2/Ormolyte/WO_3. Mol. Cryst. Liq. Cryst. 374(1): 113–118.

Balan, A., D. Baran, G. Gunbas, A. Durmus, F. Ozyurt and L. Toppare. 2009. One polymer for all: benzotriazole containing donor–acceptor type polymer as a multi-purpose material. Chem. Commun. (44): 6768–6770.

Balan, A., D. Baran and L. Toppare. 2011. Benzotriazole containing conjugated polymers for multipurpose organic electronic applications. Polym. Chem. 2(5): 1029–1043.

Balan, A., G. Gunbas, A. Durmus and L. Toppare. 2008. Donor-acceptor polymer with benzotriazole moiety: enhancing the electrochromic properties of the "donor unit". Chem. Mater. 20(24): 7510–7513.

Barbosa, P., L. Rodrigues, M. M. Silva, M. J. Smith, A. Parola, F. Pina et al. 2010. Solid-state electrochromic devices using pTMC/PEO blends as polymer electrolytes. Electrochim. Acta 55(4): 1495–1502.

Beaujuge, P. M. and J. R. Reynolds. 2010. Color control in π-conjugated organic polymers for use in electrochromic devices. Chem. Rev. 110(1): 268–320.

Blanchard, P., J. M. Raimundo and J. Roncali. 2001. New synthetic strategies towards conjugated NLO-phores and fluorophores. Synth. Met. 119(1): 527–528.

Bohnke, O., C. Bohnke, G. Robert and B. Carquille. 1982. Electrochromism in WO_3 thin films. I. $LiClO_4$-propylene carbonate-water electrolytes. Solid State Ionics 6(2): 121–128.

Bolduc, A., S. Dufresne and W. G. Skene. 2010. EDOT-containing azomethine: an easily prepared electrochromically active material with tuneable colours. J. Mater. Chem. 20(23): 4820–4826.

Bouffard, J. and T. M. Swager. 2008. Fluorescent conjugated polymers that incorporate substituted 2,1,3-benzooxadiazole and 2,1,3-benzothiadiazole units. Macromolecules 41(15): 5559–5562.

Brooke, R., E. Mitraka, S. Sardar, M. Sandberg, A. Sawatdee, M. Berggren et al. 2017. Infrared electrochromic conducting polymer devices. J. Mater. Chem. 5(23): 5824–5830.

Bryce, M. R. 1991. Recent progress on conducting organic charge-transfer salts. Chem. Soc. Rev. 20(3): 355–390.

Bryce, M. R. 2000. Functionalised tetrathiafulvalenes: new applications as versatile π-electron systems in materials chemistry. J. Mater. Chem. 10(3): 589–598.

Byker, H. J. 2001. Electrochromics and polymers. Electrochem. Acta 46(13-14): 2015–2022.

Çetin, G. A., A. Balan, A. Durmuş, G. Günbaş and L. Toppare. 2009. A new p- and n-dopable selenophene derivative and its electrochromic properties. Org. Electron. 10(1): 34–41.

Chang, C.-W., C.-H. Chung and G.-S. Liou. 2008. Novel anodic polyelectrochromic aromatic polyamides containing pendent dimethyltriphenylamine moieties. Macromolecules 41(22): 8441–8451.

Chen, L.-C. and K.-C. Ho. 2001. Design equations for complementary electrochromic devices: application to the tungsten oxide–Prussian blue system. Electrochim. Acta 46(13): 2151–2158.

Deb, S. K. 1969. A novel electrophotographic system. Appl. Opt. 8(S1): 192–195.

Deb, S. K. 1973. Optical and photoelectric properties and colour centres in thin films of tungsten oxide. Appl. Phys. 27(4): 801–822.

Ding, J., M. Day, G. Robertson and J. Roovers. 2002. Synthesis and characterization of alternating copolymers of fluorene and oxadiazole. Macromolecules 35(9): 3474–3483.

Ding, J., M. de Miguel, J. Lu, Y. Tao and H. García. 2010. Rapid switching and high contrast electrochromic property by electrochemical reduction of an alternating copolymer of fluorene and oxadiazole. J. Phys. Chem. C 114(11): 5168–5173.

Ding, P., C. Zhong, Y. Zou, C. Pan, H. Wu and Y. Cao. 2011. 5,6-Bis(decyloxy)-2,1,3-benzooxadiazole-based polymers with different electron donors for bulk-heterojunction solar cells. J. Phys. Chem. C 115(32): 16211–16219.

Durmus, A., G. E. Gunbas and L. Toppare. 2007. New, highly stable electrochromic polymers from 3,4-ethylenedioxythiophene−bis-substituted quinoxalines toward green polymeric materials. Chem. Mater. 19(25): 6247–6251.

Dyer, A. L., M. R. Craig, J. E. Babiarz, K. Kiyak and J. R. Reynolds. 2010. Orange and red to transmissive electrochromic polymers based on electron-rich dioxythiophenes. Macromolecules 43(10): 4460–4467.

Dyer, C. K. and J. S. L. Leach. 1978. Reversible optical changes within anodic oxide films on titanium and niobium. J. Electrochem. Soc. 125(1): 23.

Eh, A. L. S., A. W. M. Tan, X. Cheng, S. Magdassi and P. S. Lee. 2018. Recent advances in flexible electrochromic devices: prerequisites, challenges, and prospects. Energy Technol. 6(1): 33–45.

Ericson, H., C. Svanberg, A. Brodin, A. Grillone, S. Panero, B. Scrosati et al. 2000. Poly (methyl methacrylate)-based protonic gel electrolytes: a spectroscopic study. Electrochim. Acta 45(8-9): 1409–1414.

Göker, S., G. Hızalan, Y. A. Udum and L. Toppare. 2014. Electrochemical and optical properties of 5,6-bis(octyloxy)-2,1,3 benzooxadiazole containing low band gap polymers. Synth. Met. 191: 19–27.

Gunbas, G. E., A. Durmus and L. Toppare. 2008. A unique processable green polymer with a transmissive oxidized state for realization of potential RGB-based electrochromic device applications. Adv. Funct. Mater. 18(14): 2026–2030.

Hacioglu, S. O., S. Toksabay, M. Sendur and L. Toppare. 2014. Synthesis and electrochromic properties of triphenylamine containing copolymers: Effect of π-bridge on electrochemical properties. J. Polym. Sci. Part A: Polym. Chem. 52(4): 537–544.

Havinga, E. E., W. ten Hoeve and H. Wynberg. 1992. A new class of small band gap organic polymer conductors. Polym. Bull. 29(1): 119–126.

Hedrick, J. L. and R. Twieg. 1992. Poly(aryl ether oxadiazoles). Macromolecules 25(7): 2021–2025.

Ho, K.-C. 1999. Cycling and at-rest stabilities of a complementary electrochromic device based on tungsten oxide and Prussian blue thin films. Electrochim. Acta 44(18): 3227–3235.

Hünig, S., G. Kießlich, H. Quast and D. Scheutzow. 1973. Über zweistufige Redoxsysteme, (X1) Tetrathioäthylene und ihre höheren Oxidationsstufen. Justus Liebigs Ann. Chem. 1973(2): 310–323.

Hwang, S.-W. and Y. Chen. 2002. Photoluminescent and electrochemical properties of novel poly(aryl ether)s with isolated hole-transporting carbazole and electron-transporting 1,3,4-oxadiazole fluorophores. Macromolecules 35(14): 5438–5443.

İçli, M., M. Pamuk, F. Algi, A. M. Önal and A. Cihaner. 2010. Donor−acceptor polymer electrochromes with tunable colors and performance. Chem. Mater. 22(13): 4034–4044.

Invernale, M. A., V. Seshadri, D. M. D. Mamangun, Y. Ding, J. Filloramo and G. A. Sotzing. 2009. Polythieno [3, 4-b] thiophene as an optically transparent ion-storage layer. Chem. Mater. 21(14): 3332–3336.

Irie, M., T. Fukaminato, K. Matsuda and S. Kobatake. 2014. Photochromism of diarylethene molecules and crystals: memories, switches, and actuators. Chem. Rev. 114(24): 12174–12277.

Irvin, J. A. and J. R. Reynolds. 1998. Low-oxidation-potential conducting polymers: Alternating substituted para-phenylene and 3,4-ethylenedioxythiophene repeat units. Polymer 39(11): 2339–2347.

Jang, J., B. Lim, J. Lee and T. Hyeon. 2001. Fabrication of a novel polypyrrole/poly (methyl methacrylate) coaxial nanocable using mesoporous silica as a nanoreactor. Chem. Commun. (1): 83–84.

Jang, J. and J. H. Oh. 2005. Fabrication of a highly transparent conductive thin film from polypyrrole/poly(methyl methacrylate) core/shell nanospheres. Adv. Funct. Mater. 15(3): 494–502.

Jelle, B. P. and G. Hagen. 1999. Performance of an electrochromic window based on polyaniline, prussian blue and tungsten oxide. Sol. Energy Mater. Sol. Cells 58(3): 277–286.

Jelle, B. P., G. Hagen and S. Nødland. 1993. Transmission spectra of an electrochromic window consisting of polyaniline, prussian blue and tungsten oxide. Electrochim. Acta 38(11): 1497–1500.

Jia, P., W. A. Yee, J. Xu, C. L. Toh, J. Ma and X. Lu. 2011. Thermal stability of ionic liquid-loaded electrospun poly (vinylidene fluoride) membranes and its influences on performance of electrochromic devices. J. Membr. Sci. 376(1-2): 283–289.

Jiang, J.-M., H.-C. Chen, H.-K. Lin, C.-M. Yu, S.-C. Lan, C.-M. Liu et al. 2013. Conjugated random copolymers of benzodithiophene–benzooxadiazole–diketopyrrolopyrrole with full visible light absorption for bulk heterojunction solar cells. Polym. Chem. 4(20): 5321–5328.

Jiang, J.-M., P.-A. Yang, T.-H. Hsieh and K.-H. Wei. 2011. Crystalline low-band gap polymers comprising thiophene and 2,1,3-benzooxadiazole units for bulk heterojunction solar cells. Macromolecules 44(23): 9155–9163.

Jin, A., W. Chen, Q. Zhu, Y. Yang, V. L. Volkov and G. S. Zakharova. 2009. Structural and electrochromic properties of molybdenum doped vanadium pentoxide thin films by sol–gel and hydrothermal synthesis. Thin Solid Films 517(6): 2023–2028.

Joseph, J., H. Gomathi and G. P. Rao. 1991. Electrodes modified with cobalt hexacyanoferrate. J. Electroanal. Chem. Interfacial Electrochem. 304(1): 263–269.

Kamtekar, K. T., H. L. Vaughan, B. P. Lyons, A. P. Monkman, S. U. Pandya and M. R. Bryce. 2010. Synthesis and Spectroscopy of Poly(9,9-dioctylfluorene-2,7-diyl-co-2,8-dihexyldibenzothiophene-S,S-dioxide-3,7-diyl)s: Solution-processable, deep-blue emitters with a high triplet energy. Macromolecules 43(10): 4481–4488.

Karsten, B. P., L. Viani, J. Gierschner, J. Cornil and R. A. J. Janssen. 2009. On the origin of small band gaps in alternating thiophene−thienopyrazine oligomers. J. Phys. Chem. A 113(38): 10343–10350.

Kaya, E., A. Balan, D. Baran, A. Cirpan and L. Toppare. 2011. Electrochromic and optical studies of solution processable benzotriazole and fluorene containing copolymers. Org. Electron. 12(1): 202–209.

Kozma, E., F. Munno, D. Kotowski, F. Bertini, S. Luzzati and M. Catellani. 2010. Synthesis and characterization of perylene-based donor–acceptor copolymers containing triple bonds. Synth. Met. 160(9): 996–1001.

Kumar, A., D. M. Welsh, M. C. Morvant, F. Piroux, K. A. Abboud and J. R. Reynolds. 1998. Conducting poly(3,4-alkylenedioxythiophene) derivatives as fast electrochromics with high-contrast ratios. Chem. Mater. 10(3): 896–902.

Kung, Y.-C. and S.-H. Hsiao. 2011. Novel luminescent and electrochromic polyhydrazides and polyoxadiazoles bearing pyrenylamine moieties. Polym. Chem. 2(8): 1720–1727.

Kuo, C.-W., J.-C. Chang, L.-T. Lee, J.-K. Chang, Y.-T. Huang, P.-Y. Lee et al. 2022. Electrosynthesis of electrochromic polymers based on bis-(4-(N-carbazolyl)phenyl)-phenylphosphine oxide and 3,4-propylenedioxythiophene derivatives and studies of their applications in high contrast dual type electrochromic devices. J. Taiwan Inst. Chem. Eng. 131: 104173.

Li, W., R. Qin, Y. Zhou, M. Andersson, F. Li, C. Zhang et al. 2010. Tailoring side chains of low band gap polymers for high efficiency polymer solar cells. Polymer 51(14): 3031–3038.

Lin, H.-Y., G.-S. Liou, W.-Y. Lee and W.-C. Chen. 2007. Poly(triarylamine): Its synthesis, properties, and blend with polyfluorene for white-light electroluminescence. J. Polym. Sci. Part A: Polym. Chem. 45(9): 1727–1736.

Liou, G.-S., S.-H. Hsiao, H.-M. Huang, C.-W. Chang and H.-J. Yen. 2007. Synthesis and photophysical properties of novel organo-soluble polyarylates bearing triphenylamine moieties. J. Polym. Res. 14(3): 191–199.

Liou, G.-S., S.-H. Hsiao and T.-H. Su. 2005. Novel thermally stable poly(amine hydrazide)s and poly(amine-1,3,4-oxadiazole)s for luminescent and electrochromic materials. J. Polym. Sci., Part A: Polym. Chem. 43: 3245–3256.

Luo, Z., J. Yang, H. Cai, H. Li, X. Ren, J. Liu et al. 2008. Preparation of silane-WO_3 film through sol–gel method and characterization of photochromism. Thin Solid Films 516(16): 5541–5544.

Marcilla, R., F. Alcaide, H. Sardon, J. A. Pomposo, C. Pozo-Gonzalo and D. Mecerreyes. 2006. Tailor-made polymer electrolytes based upon ionic liquids and their application in all-plastic electrochromic devices. Electrochem. Commun. 8(3): 482–488.

McQuade, D. T., A. E. Pullen and T. M. Swager. 2000. Conjugated polymer-based chemical sensors. Chem. Rev. 100(7): 2537–2574.

Mishra, S. P., K. Krishnamoorthy, R. Sahoo and A. Kumar. 2005. Synthesis and characterization of monosubstituted and disubstituted poly (3, 4-propylenedioxythiophene) derivatives with high electrochromic contrast in the visible region. J. Polym. Sci. A Polym. Chem. 43(2): 419–428.

Möller, M., S. Asaftei, D. Corr, M. Ryan and L. Walder. 2004. Switchable electrochromic images based on a combined top–down bottom–up approach. Adv. Mater. 16(17): 1558–1562.

Mondal, R., H. A. Becerril, E. Verploegen, D. Kim, J. E. Norton, S. Ko et al. 2010. Thiophene-rich fused-aromatic thienopyrazine acceptor for donor–acceptor low band-gap polymers for OTFT and polymer solar cell applications. J. Mater. Chem. 20(28): 5823–5834.

Mortimer, R. J. 1997. Electrochromic materials. Chem. Soc Rev. 26(3): 147–156.

Mortimer, R. J. 1999. Organic electrochromic materials. Electrochim. Acta 44(18): 2971–2981.

Moser, M. L., G. Li, M. Chen, E. Bekyarova, M. E. Itkis and R. C. Haddon. 2016. Fast electrochromic device based on single-walled carbon nanotube thin films. Nano Lett. 16(9): 5386–5393.

Munro, B., P. Conrad, S. Krämer, H. Schmidt and P. Zapp. 1998. Development of electrochromic cells by the sol–gel process. Sol. Energy Mater. Sol. Cells 54(1-4): 131–137.

Niklasson, G. A. and C. G. Granqvist. 2007. Electrochromics for smart windows: thin films of tungsten oxide and nickel oxide, and devices based on these. J. Mater. Chem. 17(2): 127–156.

Nunes, S., V. de Zea Bermudez, D. Ostrovskii, P. Barbosa, M. M. Silva and M. J. Smith. 2008. Cationic and anionic environments in LiTFSI-doped di-ureasils with application in solid-state electrochromic devices. Chem. Phys. 345(1): 32–40.

Oßwald, S., S. Breimaier, M. Linseis and R. F. Winter. 2017. Polyelectrochromic vinyl ruthenium-modified tritylium dyes. Organometallics 36(10): 1993–2003.

Ozelcaglayan, A. C., M. Sendur, N. Akbasoglu, D. H. Apaydin, A. Cirpan and L. Toppare. 2012. Synthesis and electrochemical properties of a new benzimidazole derivative as the acceptor unit in donor–acceptor–donor type polymers. Electrochim. Acta 67: 224–229.

Özkut, M. İ., M. P. Algi, Z. Öztaş, F. Algi, A. M. Önal and A. Cihaner. 2012. Members of CMY color space: cyan and magenta colored polymers based on oxadiazole acceptor unit. Macromolecules 45(2): 729–734.

Palenzuela, J., A. Viñuales, I. Odriozola, G. Cabañero, H. J. Grande and V. Ruiz. 2014. Flexible viologen electrochromic devices with low operational voltages using reduced graphene oxide electrodes. ACS Appl. Mater. Interfaces 6(16): 14562–14567.

Patel, K., G. Bhatt, J. Ray, P. Suryavanshi and C. Panchal. 2017. All-inorganic solid-state electrochromic devices: a review. J. Solid State Electrochem. 21: 337–347.

Puguan, J. M. C., W.-J. Chung and H. Kim. 2016. Ion-conductive and transparent PVdF-HFP/silane-functionalized ZrO_2 nanocomposite electrolyte for electrochromic applications. Electrochim. Acta 196: 236–244.

Rani, V. and K. S. V. Santhanam. 1998. Polycarbazole-based electrochemical transistor. J. Solid State Electr. 2(2): 99–101.

Rosseinsky, D. R. and R. J. Mortimer. 2001. Electrochromic systems and the prospects for devices. Adv. Mater. 13(11): 783–793.

Seeboth, A., D. Lötzsch, R. Ruhmann and O. Muehling. 2014. Thermochromic polymers—function by design. Chem. Rev. 114(5): 3037–3068.

Segura, J. L. and N. Martín. 2001. New concepts in tetrathiafulvalene chemistry. Angew. Chem., Int. Ed. 40(8): 1372–1409.

Seo, J. H., A. Gutacker, Y. Sun, H. Wu, F. Huang, Y. Cao et al. 2011. Improved high-efficiency organic solar cells via incorporation of a conjugated polyelectrolyte interlayer. J. Am. Chem. Soc. 133(22): 8416–8419.

Sotzing, G. A., J. R. Reynolds and P. J. Steel. 1996. Electrochromic conducting polymers via electrochemical polymerization of bis(2-(3,4-ethylenedioxy)thienyl) monomers. Chem. Mater. 8(4): 882–889.

Talagaeva, N. V., E. V. Zolotukhina, P. A. Pisareva and M. A. Vorotyntsev. 2016. Electrochromic properties of Prussian blue–polypyrrole composite films in dependence on parameters of synthetic procedure. J. Solid State Electrochem. 20(5): 1235–1240.

Tang, T., T. Lin, F. Wang and C. He. 2014. Azulene-based conjugated polymers with tuneable near-IR absorption up to 2.5 μm. Polym. Chem. 5(8): 2980–2989.

Tao, X.-T., Y.-D. Zhang, T. Wada, H. Sasabe, H. Suzuki, T. Watanabe et al. 1998. Hyperbranched polymers for electroluminescence applications. Adv. Mater. 10(3): 226–230.

Thakur, V. K., G. Ding, J. Ma, P. S. Lee and X. Lu. 2012. Hybrid materials and polymer electrolytes for electrochromic device applications. Adv. Mater. 24(30): 4071–4096.

Thomas, C. A., K. Zong, K. A. Abboud, P. J. Steel and J. R. Reynolds. 2004. Donor-mediated band gap reduction in a homologous series of conjugated polymers. J. Am. Chem. Soc. 126(50): 16440–16450.

Udum, Y. A., C. G. Hızlıateş, Y. Ergün and L. Toppare. 2015. Electrosynthesis and characterization of an electrochromic material containing biscarbazole–oxadiazole units and its application in an electrochromic device. Thin Solid Films 595: 61–67.

Wang, C., A. S. Batsanov and M. R. Bryce. 2004. Electrochromic tetrathiafulvalene derivatives functionalised with 2,5-diaryl-1,3,4-oxadiazole chromophores. Chem. Commun. (5): 578–579.

Wang, H., H. Huang and S. L. Wunder. 2000. Novel microporous poly (vinylidene fluoride) blend electrolytes for lithium-ion batteries. J. Electrochem. Soc. 147(8): 2853.

Wen, R.-T., G. A. Niklasson and C. G. Granqvist. 2014. Electrochromic nickel oxide films and their compatibility with potassium hydroxide and lithium perchlorate in propylene carbonate: Optical, electrochemical and stress-related properties. Thin Solid Films 565: 128–135.

Wu, F.-I., P.-I. Shih, C.-F. Shu, Y.-L. Tung and Y. Chi. 2005. Highly efficient light-emitting diodes based on fluorene copolymer consisting of triarylamine units in the main chain and oxadiazole pendent groups. Macromolecules 38(22): 9028–9036.

Wu, T.-Y., J.-C. Chang, Y.-C. Lin, J.-E. Chiang, C.-H. Yeh, L.-T. Lee et al. 2023. Synthesis and multicolored electrochromism of polycarbazoles containing oxadiazole. Dyes Pigm. 213: 111157.

Wudl, F., G. M. Smith and E. J. Hufnagel. 1970. Bis-1,3-dithiolium chloride: an unusually stable organic radical cation. J. Chem. Soc., Chem. Commun. (21): 1453–1454.

Yan, R., L. Liu, H. Zhao, Y. G. Zhu, C. Jia, M. Han et al. 2016. A TCO-free prussian blue-based redox-flow electrochromic window. J. Mater. Chem. C 4(38): 8997–9002.

Yang, P., P. Sun and W. Mai. 2016. Electrochromic energy storage devices. Mater. Today 19(7): 394–402.

Yen, H.-J. and G.-S. Liou. 2009. Synthesis, photoluminescence, and electrochromism of novel aromatic poly(amine-1,3,4-oxadiazole)s bearing anthrylamine chromophores. J. Polym. Sci., Part A: Polym. Chem. 47(6): 1584–1594.

Zaifoglu, B., M. Sendur, N. A. Unlu and L. Toppare. 2012. High optical transparency in all redox states: Synthesis and characterization of benzimidazole and thieno[3,2-b]thiophene containing conjugated polymers. Electrochim. Acta 85: 78–83.

Zhang, D., M. Wang, X. Liu and J. Zhao. 2016. Synthesis and characterization of donor–acceptor type conducting polymers containing benzotriazole acceptor and benzodithiophene donor or s-indacenodithiophene donor. RSC Adv. 6(96): 94014–94023.

Zhou, D., R. Zhou, C. Chen, W.-A. Yee, J. Kong, G. Ding et al. 2013. Non-volatile polymer electrolyte based on poly (propylene carbonate), ionic liquid, and lithium perchlorate for electrochromic devices. J. Phys. Chem. B 117(25): 7783–7789.

Chapter 5

Oxadiazole as Nonlinear Optical (NLO) Materials

1. Introduction

Compounds comprising chromophoric groups in combination with aromatic heterocyclic rings exhibit significant optical output, making them relevant for their potential application in the construction of nonlinear optical materials (Leng et al. 2001; Zadrożna and Kaczorowska 2006; Homocianu et al. 2019). Changes in the optical parameters, like absorption coefficient and index of refraction due to increasing input light intensity, culminated in the establishment of the nonlinear optical phenomena—second harmonic generation (SHG)—that is discernible only when the laser was improved in 1962 (Nie 1993; Maria 2018). Consequently, nonlinear optics grew into a massive field of study, especially after a thorough understanding of nonlinear optics phenomena (NLO) and the structure-property relationships of chromophores, as well as the development of various tools for accurately measuring and calculating hyperpolarizabilities (Caricato et al. 2018; Maria 2018). Recent literature emphasizes the growing interest in organic compounds over the last few decades as a viable alternative to their inorganic analogs, with numerous benefits, such as inexpensiveness, low environmental impact, ease of the solution processability, versatility for device fabrications (Mukherjee and Thilagar 2014), and modulation of their optical, electronic, and chemical properties by modifying their molecular structure. Indoor light sources include field effect transistors, photovoltaic devices, organic light-emitting diodes (OLEDs), and white light sources (Maria 2018).

Owing to the potential uses of nonlinear optical (NLO) properties of organic materials in optical communication, optical data storage, photodynamic therapy, three-dimensional memory, and photonic devices, such as optical switches, the organic materials with NLO properties have attracted a lot of attention (Frederiksen et al. 2005; Duan et al. 2016; Homocianu et al. 2019). The convenient processing of organic compounds into thin films, a process essential for constructing novel, successful devices, involves precisely monitoring their chemical, structural, and morphological properties (Chopra et al. 2004; Maria 2018). Organic substance deposition in thin films must meet market requirements, such as (1) a high degree of uniformity of simple or multilayered structures of organic, polymeric, or composite materials—in the electronics sector; (2) thickness management, film consistency during deposition with desirable interfacing properties—in OLED polymer uses; and (3) conformal coatings necessary to alter the interior surfaces of porous materials (membranes, foams, and textiles) or irregular geometries (Baxamusa 2015; Maria 2018).

Organic materials having a heterocyclic ring conjugated with a chromophoric group display improved donor-acceptor capability, which has a beneficial influence on the NLO response of the compound (Li et al. 2006; Morley and Whittaker 2006; Homocianu et al. 2019). Moreover, the systems having a strongly polarized and noncentrosymmetrical chemical moiety with donor and acceptor substituents, as well as a strong intramolecular charge transfer, also positively influence the NLO response (Qiu et al. 2013; Homocianu et al. 2019). Due to strong charge transfer in these systems, a large number of heterocyclic organic compounds like benzothiazoles, oxadiazoles, oligothiophenes, thienylphthalazines, azobenzenes, thienylpyrroles, and the majority of coumarin derivatives have been extensively investigated as an efficient NLO material, and they have been found to exhibit high nonlinear optical susceptibilities when compared to other materials (Chun et al. 2001; Kumar et al. 2016; Tathe and Sekar 2016).

Among them, oxadiazoles are excellent candidates in materials science as NLO materials and organic light-emitting diodes because of their electron-deficient character (Wang et al. 1999). Oxadiazoles belong to the azole family of five-membered heterocyclic compounds having three possible isomeric forms. Out of four possible isomeric forms of oxadiazole, namely 1,2,4-, 1,3,5-, 1,2,3- and 1,3,4-, three isomeric forms, namely 1,3,4-, 1,2,4- and 1,3,5- are stable, while 1,2,3-isomeric form isomerizes to diazoketone tautomeric form (Joule and Mills 2012). Among them, a 1,3,4-isomeric form of oxadiazole, due to its greater stability and easy functionalization through nucleophilic substitution at the thiol group, has been extensively investigated for various applications. Therefore, the features and practical synthetic processes of highly effective materials manufacturing NLO materials have been extensively investigated and reported. Since it was discovered that oxadiazoles exhibit the necessary structural characteristics to function as effective electron transporters, multiple research teams have concentrated on the synthesis of various small molecules and polymers containing this heterocyclic moiety. The oxadiazole derivatives were found to be excellent candidates for creating electroluminescent devices after their optical and electrical properties were characterized. With the advent of metal-catalyzed cross-coupling processes, the chemical synthesis of compounds with a range of heterocyclic cores and sensible organic groups has significantly changed.

The following section describes the various donor-acceptor combinations with oxadiazole as one component for developing material for application as an NLO device.

2. Oxadiazole Derivatives With Donor-Acceptor Properties

Oxadiazole derivatives have been used as a donor p-electron relay to develop a variety of donor-acceptor molecules with varied applications, including NLO materials. Such donor-acceptor molecules have the potential to exhibit excellent nonlinear properties. In reported design strategies, the successful development of the D-A molecular system requires connecting a donor and an acceptor group at the terminal via a p-bridge, which allows the creation of a polarized system to achieve large molecular nonlinearity. Various designs have appeared in literature, which included D-A olefins (Marder et al. 1994; Blanchard-Desce et al. 1995), acetylenes (Cheng et al. 1991), aromatic (Cheng et al. 1991), heteroaromatic rings (Rao et al. 1993; Rao et al. 1994) and azo moieties (Moylan et al. 1993) as p-bridges to obtain efficient NLO materials (Mashraqui et al. 2004). However, push-pull systems developed using polyenes are known for poor thermal stability despite having large first hyperpolarizability β, while aromatic D-A systems display good stability but poor values of β. Therefore, oxadiazole-containing NLO materials have been developed, with a polarizable five-membered ring as a p-bridge in D-A molecular systems. A variety of donors and acceptors have been employed to tune the electron delocalization and improve second-order nonlinearity. In view of the above-stated facts, Mashraqui et al. (2004) have developed a variety of D-A acceptor systems with 1,3-4-oxadiazole moiety as p-bridge (1,2,3a–f). The thermal stability of the molecular system were studied using thermal scanning calorimetry and obtained their decomposition temperature listed in Table 1. The molecular systems (1,2,3a–f) displayed good stability by showing a more than 200°C thermal decomposition (Td) temperature. The Td of 3d was not reported due to its decomposition accompanied with melting at a temperature of more 410°C.

The UV-vis data and first hyperpolarizability β of the oxadiazole compounds were reported using the hyper-Rayleigh Scattering technique (Clays and Persoons 1991; Krishnan et al. 2001) at 1,064 nm using external reference (Table 1). No correction in molecular hyperpolarizability was performed for two-photon fluorescence.

In the absorption profile for 1a, an absorption band at 338 nm was witnessed, which shifted bathochromically by 82 nm in highly polarized 1c (420 nm) due to the

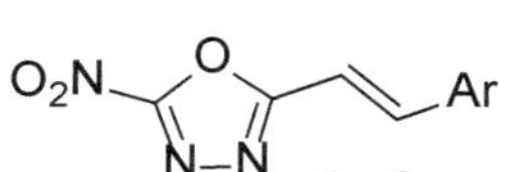

1a-f

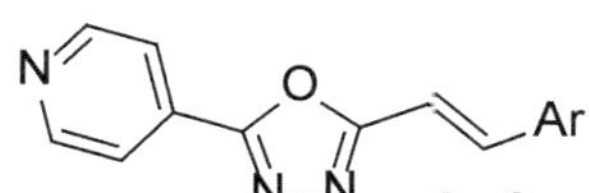

2a-f

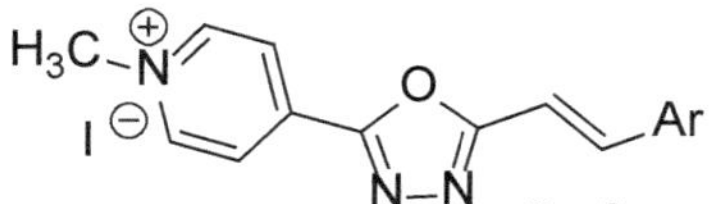

3a-f

1a; 2a; 3a; Ar = Ph-
1b; 2b; 3b; Ar = 4-MeO-Ph-
1c; 2c; 3c; Ar = 4-Me2N-Ph-
1d; 2d; 3d; Ar = 2-furyl-
1e; 2e; 3e; Ar = 2-thienyl-
1f; 2f; 3f; Ar = 1-Methyl-2-pyrrolyl-

Structures 1–3a–f

presence of a dimethylaniline (strong electron donor) in the structure of 1c. A steady increase in the value of β was witnessed from 1a to 1c, which may be attributed to a bathochromic shift in the absorption band due to the dispersion characteristic of second-order nonlinearity. In oxadiazoles 1d, 1e, and 1f, which have intermediate strength p-donor chromophores, the charge transfer bond is witnessed in the 338–420 nm range. The UV-vis spectrophotometric data and first hyperpolarizability β of various oxadiazole derivatives were obtained using the hyper-Rayleigh Scattering methodology at 1,064 nm in solution state as per the established method (Table 1) (Clays and Persoons 1991; Krishnan et al. 2001). Equation 1 was used for the calculation of β_0 using a state model. The term "ω" represents the laser fundamental, while "ω_0" represents the single photon absorption maximum observed for the molecular (in wavenumber).

$$\frac{\beta}{\beta_0} = \frac{(\omega)^4}{[(\omega_0^2 - \omega^2)(\omega_0^2 - 4\omega^2)]} \qquad \text{(Eq. 1)}$$

Among 1a–f, based on relative donor strength, the lowest β_o was observed for phenyl oxadiazole 1b and 1c (Table 1), while 1e and 1f displayed higher βo values (Table 1). Moreover, the oxadiazoles 1d–f containing aromatic heterocyclic rings with lower resonance stability energies as compared to benzene were not observed to be conducive to enhancing the dispersion-free hyperpolarizability. Therefore, it can be concluded that oxadiazole-containing donors like -NMe_2 or -OMe should be favored over others for improving second-order nonlinearity. The charge transfer bands in 2a–f were witnessed in the 317 nm to 400 nm range, which shifted bathochromically by 36–59 nm in the corresponding oxadiazole salts 3a–f (353–400 nm, Table 1). The experimental data was attributed to an increased p-electron acceptor ability of the pyridinium ring compared to pyridine.

Further, the bathochromic shift in the absorption band was accompanied by an increase in values of β (Mashraqui et al. 2004). The study demonstrated that D-A oxadiazole systems containing strong donors like p-methoxyphenyl/p-dimethylaminophenol and acceptors like pyridinium exhibit large second-order nonlinearity. Further, it was demonstrated the oxadiazole core could function as an effective p-bride in the design of NLO materials. However,

Table 1. Electronic absorption and hyperpolarizability of D–A oxadiazoles 1a–f, 2a–f and 3a–f. Reproduced here with the permission of Elsevier (Mashraqui et al. 2004).

Comp.	λ_{max} nm	$\beta \times 10^{-30}$[a] esu	$\beta_0 \times 10^{-30}$ esu	T_d °C	Comp.	λ_{max} nm	$\beta \times 10^{-30}$ esu	$\beta_0 \times 10^{-30}$ esu	T_d °C	Comp.	λ_{max} nm	$\beta \times 10^{-30}$ esu	$\beta_0 \times 10^{-30}$ esu	T_d °C
1a	338	34.0	18.3	248	**2a**	317	46.0	27.2	212	**3a**	353	77.0	38.5	248
1b	360	49.0	23.5	204	**2b**	340	50.0	28.0	–	**3b**	382	94.7	39.5	206
1c	420	60.1	19.1	203	**2c**	400	60.0	23.0	279	**3c**	459	123.6	25.7	246
1d	357	39.7	19.5	223	**2d**	341	14.1	7.8	199	**3d**	377	59.6	26.0	>410
1e	358	32.6	15.8	–	**2e**	343	28.1	14.8	223	**3e**	374	57.3	26.0	200
1f	390	38.1	15.2	177	**2f**	374	28.0	12.7	254	**3f**	430	–	–	226

[a] In the external reference method the measured β values were calibrated against paranitroaniline in chloroform and in some cases methanol as the standard. The β values used for paranitroaniline are 17.8×10^{-30} esu and 22.0×10^{-30} esu in chloroform and methanol, respectively.

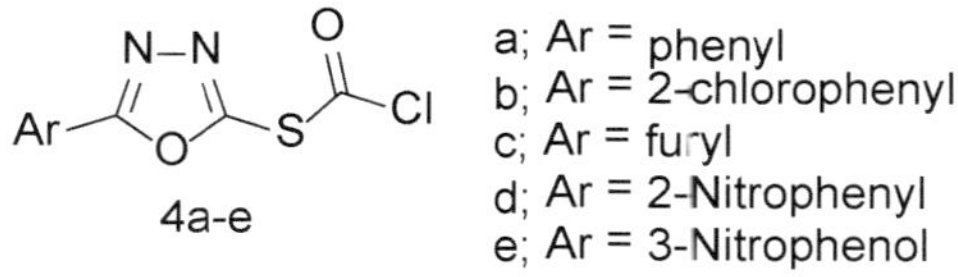

Structure 4a–e

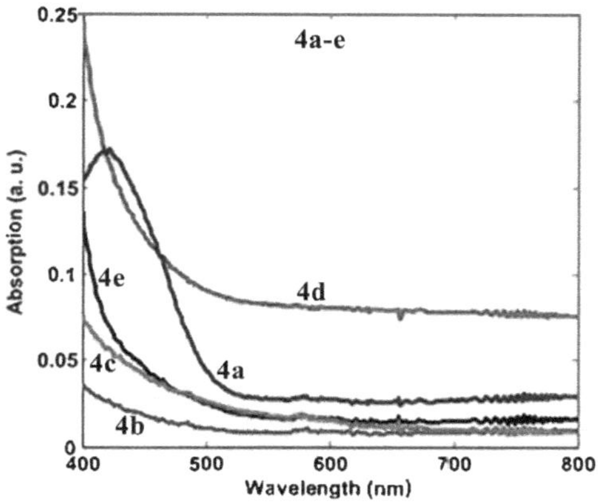

Figure 1. UV-vis spectra of 4a–e obtained in dichloromethane (Ghezelbash et al. 2019).

heterocycles like furan and thiophene have low stabilization energy compared to benzene and do not significantly alter hyperpolarizability (Mashraqui et al. 2004).

Ghezelbash et al. have synthesized five S-(5-aryl-1,3,4-oxadiazol-2-yl)2-chloroethanethioate (4a–e) derivatives and investigated their NLO properties (Ghezelbash et al. 2019). UV-visible data provided the 0.028, 0.008, 0.020, 0.084, and 0.018 absorbance values at 532 nm for 4a–e (Figure 1). The linear absorption coefficient (α) can be determined using equation α = (1/L)LηT, where "L" is the path length, "T" is the temperature, and "η" represents the refractive index. The nonlinear refractive indexes of 4a–e were recorded in dichloromethane using Z-scan methods with a continuous diode-pumped laser at 532 nm (Figure 2).

Figure 2 shows the closed aperture Z-scan of samples of 4a–e using the same incident laser intensity and linear transmittance of 0.1. The area of minimum transmittance (valley) before focus and maximum transmittance (Peak) after focus was witnessed due to the self-defocusing formation. This effect was represented by a negative nonlinear refractive coefficient (n_2). Therefore, features in the graph reveal the sign of the refractive index. The continuous wave laser caused an increase in the temperature (ΔT) at the laser spot, causing a change in the refractive index. Therefore, Δn = (dn/dT)ΔT, where "dn/dT" represents the thermo-optic coefficient of the sample and can be calculated from the density and temperature changes of the sample. The peak and value differences (ΔT_{pv}) and changes in phase shift can be calculated using Equations 2 and 3.

$$\Delta T_{pv} = 0.0406\,(IS)^{0.25}|\,\Delta\phi_0| \qquad \text{(Eq. 2)}$$

$$\Delta\phi = kn_2 I_0 L_{eff} \qquad \text{(Eq. 3)}$$

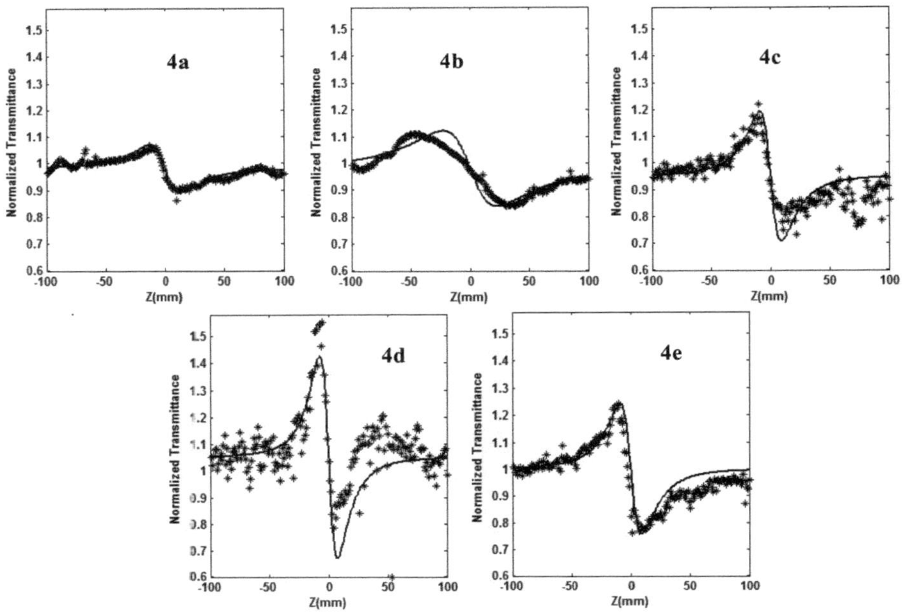

Figure 2. Closed aperture curves of five compounds 4a–e in dichloromethane at I_0 = 89.36 kW/m^2 incident intensity of laser beam (Ghezelbash et al. 2019).

Table 2. Measured values of Linear absorption coefficient (α), ΔT_{pv}, η_2 of 4a–e at I_0 = 89.36 kW/m^2 (Ghezelbash et al. 2019).

Sample	$\alpha\ (cm^1)$	ΔT_{PV}	$n_2(\times 10^{11}\ m^2/W)$
(**a**) Ar = phenyl	0.065	0.170	-6.08
(**b**) 2-Chlorophenyl	0.020	0.268	-9.89
(**c**) Ar = Furyl	0.047	0.471	-17.35
(**d**) 2-Nitrophenyl	0.193	0.767	-29.21
(**e**) 3-Nitrophenyl	0.041	0.470	-17.39

The term "S" in Equation 2 represents the linear transmittance of the aperture, which is about 0.1, and "$\Delta\phi_0$" represents changes in the phase. The term "*k*" in Equation 3 represents the wave vector, which can be calculated using the term "$k = 2\pi/\lambda$", "Io" represents the incident intensity of the laser, and Leff is the effective sample length ($L_{eff} = 1e^{\alpha l}/\alpha$). The NLO properties of compounds 4a–e are listed in Table 2.

It was reported that the NLO values of compounds 4a–e increased with an increase in the electron-withdrawing effects of the substituents. The compound 4d provided a better NLO property than compound 4e due to the presence of the ortho nitro group, which leads to better resonance contributing forms and delocalization. The electron-withdrawing effect of the furyl substituent in 4c is more than the 2-chlorophenol in 4b due to the presence of a more electronegative oxygen atom in the furyl ring. As expected, the phenyl ring is less electronegative than the chlorophenyl group.

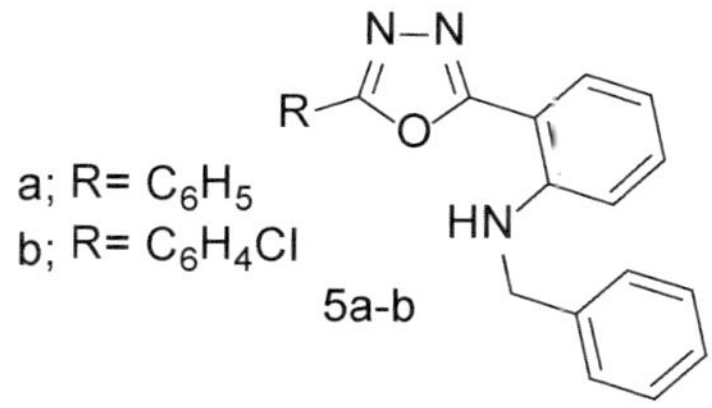

Structures 5a–b

Dhonnar et al. have recently reported NLO properties of two 1,3,4-oxadiazole derivatives 5a,b (Dhonnar et al. 2022). However, it was observed that both 5a and b display poor nonlinear optical properties.

2.1 Polyacetylenes Containing Oxadiazole Pendant Groups

Polytypical conjugated polymers like polyacetylene exhibit strong nonlinear optical properties and fast response time, which led to their application in a variety of photonic devices (Shi et al. 2000; Lee et al. 2002; Liu et al. 2009; Wang et al. 2010). The significantly strong *p*-electron delocalization in the polyacetylene (PA) chain generates a large molecule polarizability and, as a result, noteworthy third-order optical nonlinearities (Fann et al. 1989). However, low stability and intractability limit the applications of polyacetylene (Liu et al. 2009). The problems associated with polyacetylene can be improved by introducing suitable substituents in its backbone to impart superior properties to the material (Lam and Tang 2005; Liu et al. 2009; Wang et al. 2010). For instance, the introduction of chromophoric functional pendants like naphthalene, anthracene, carbazole, azobenzene, stilbene, and oxadiazole into the polyacetylene chain led to dramatic improvement in the third-order nonlinear optical susceptibilities and solubilities (Sata et al. 1998; Nanjo et al. 1999; Nomura et al. 2000; Yin et al. 2007; Su et al. 2008; Wang et al. 2008a; Wang et al. 2008b; Wang et al. 2010). Among various segments, which can be appended to improve the properties of polyacetylene as an NLO material, oxadiazoles derivatives represent interesting chromophoric groups with electron-withdrawing properties, weak ground state electronic absorption in the visible region above 400 nm (He et al. 2004; Jin et al. 2004; Wang et al. 2008a) and good thermal stability (Sung and Lin 2004) along with large second-order nonlinear susceptibility (Mashraqui et al. 2004; Wang et al. 2008b). As per Schuling's nonlinear optical theory, the third-order nonlinear optical susceptibility (γ) is dependent on γ_e^0, which represents the delocalization of electrons and β (second-order nonlinear optical susceptibility) (Schweig 1967).

Based on the above-stated facts, Wang et al. have designed solvent-dissolvable polyacetylenes having oxadiazole segments within the main polymeric chains (Scheme 1) (Wang et al. 2008b).

It was reported that polyacetylene polymers having oxadiazole units with flexible chains displayed greater solubility. Therefore, the polymer PO1 displayed greater solubility than PO2. The polymer PO2 displayed partial solubility in solvents like THF, $CHCl_3$, toluene, and dioxane, which was better than polymers having no alkoxy group at the terminal and provided a design principle for building a suitably functionalized polyacetylenes for NLO applications. The good thermal stability of

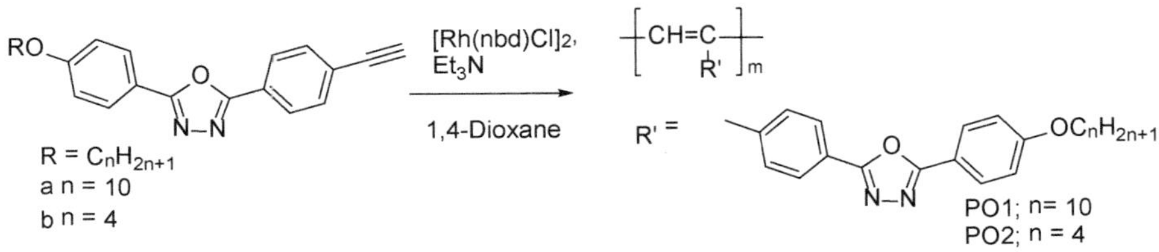

Scheme 1. Synthesis of polyacetylene polymer containing oxadiazole unit in the main chain.

both polymers was witnessed based on thermogravimetric analysis (TGA). In the case of PO1, the thermal degradation (5% weight loss) was witnessed at 353°C, while for PO2, it occurred at 346°C. The stability may be ascribed to the formation of a jacket of aromatic oxadiazole around the polymer chain through strong intermolecular interactions, which shielded the main chains from decomposition due to increase in temperature (Lam et al. 2002; Yin et al. 2005; Wang et al. 2008).

Moreover, the stereoregularity of the polyacetylene chain also affected the thermal stability. An increase in the *cis* olefin content of the polymer increases the thermal stability, which can be adjusted by the alkoxy chain (Tang et al. 1997; Wang et al. 2008b). The optical limiting performance of PO1 and PO2 were compared with poly(phenyleneacetylene) (PPA) at a transmittance value of (67%) in THF using a 532 nm laser. At low input fluence, the output fluence increased linearly with input fluence. However, as the input fluence is increased further, a decrease in the transmittance of PO1 and PO2 solution was witnessed, and a nonlinear relation between input and output fluence was observed, displaying substantial optical limiting properties (Figure 3). It should be noted from Figure 3 that a PPA solution displayed a continuous increase in the transmittance owing to laser-induced photolysis of polyacetylene chains. The experiment demonstrated that introducing oxadiazole moiety in polyacetylene chains imparted its nonlinear optical properties. To check the photostability of PO1 and PO2, UV-vis spectra before and after

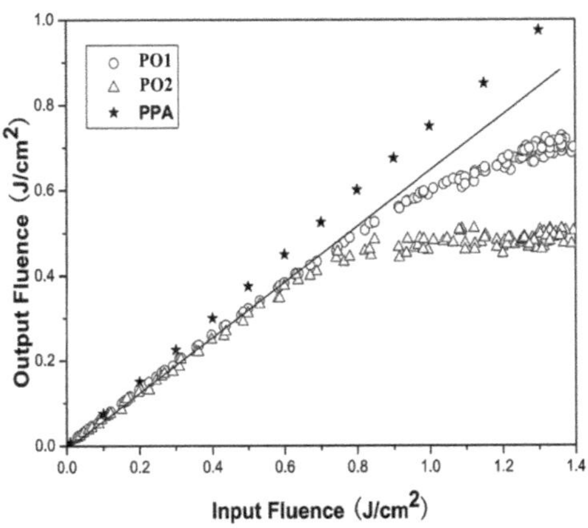

Figure 3. Optical limiting properties of PO1 and PO2 solution (PO1: Table 1, run 8 and PO2: Table 2, run 8) with a linear transmission of 67%. Data for PPA with the same transmission are given for comparison. Reproduced here with the permission of RSC (Wang et al. 2008).

Table 3. Summary of optical limiting and nonlinear optical properties of PO1 (Table 1, run 8) and PO2 (run 8). Reproduced here with the permission of RSC (Wang et al. 2008a).

	OL Amplitude $(J/cm^2)^a$	NLO Properties[b]		
		β ($\times 10^{-11}$m/W)	n_2 ($\times 10^{-19}m^2/W$)	$\chi^{(3)}$ ($\times 10^{-12}$esu)
PO1	0.454	8.54	16.10	4.67
PO2	0.713	4.75	6.94	2.15

[a] Maximum output intensity. [b] Measured by the Z-scan technique with an 13-ns Nd:YAG laser system at a 1-Hz repetition rate and a 532-nm wavelength.

photolysis was recorded, which displayed no significant changes in the absorption profile, suggesting good stability under radiation (Wang et al. 2008b). Moreover, it was observed that at an input fluence of 0.625 J/cm² (called limiting threshold), the solution of PO1 started to show nonlinearity.

Here, the limiting threshold is defined as the input fluence at which the output fluence begins to show nonlinearity. With a further increase in input fluence, the output fluence reached a plateau and showed saturation at 0.454 J/cm² (known as limiting amplitude). Here, the limit amplitude may be defined as the maximum output intensity. For the PO2 solution, the limiting value of amplitude was observed at 0.713 J/cm². The observation suggested that PO1, with a longer alkoxy chain and greater cis olefin content, displays better optical limiting performance as compared to PO2, with a shorter alkoxy chain at the terminal and lower cis olefin content (Table 3) (Wang et al. 2008a). Authors have utilized a wavelength of 532 nm with 4 ns pulse width and observed a decrease in the transmittance of the solution with an increase in input fluence, which indicated that the optical limiting properties in PO1 and PO2 may have originated from reverse saturable absorption (RSA). The nonlinear absorption coefficients of polyacetylene polymers were calculated using the Z-scan technique (Figure 4). The normalized transmittance for the open aperture was expressed based on Equations 4–6. The term "$\alpha 2$" represents the nonlinear absorption coefficient, "$I_o(t)$" represents the laser light intensity at z = 0, "α_o" is the linear absorption coefficient, "L" is the thickness of the sample, "z_0" represents the diffraction length of the beam, and "z" represents the sample position (Sheik-Bahae et al. 1990).

$$T(z,s=1)=\sum_{m=0}^{\infty}\frac{[-q_0(z,0)]^m}{(m+1)^{3/2}} \text{ for } |q_0|<1 \quad \text{(Eq. 4)}$$

$$q_0(z) = \alpha_0 I_{(t)} L_{eff}(1 + Z^2/X_o^2) \quad \text{(Eq. 5)}$$

$$L_{eff} = |1 - \exp(-\alpha_0 L)|/\alpha_0 \quad \text{(Eq. 6)}$$

The normalized transmittance for the closed aperture Z-scan is represented by Equations 7 and 8.

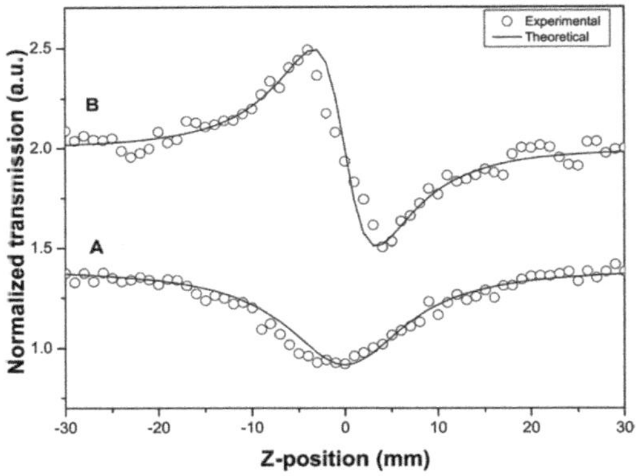

Figure 4. Z-scan data of (A) open and (B) closed apertures of PO1 (run 8). Reproduced here with the permission of RSC (Wang et al. 2008a).

$$T(z) = 1 \frac{4\Delta\phi_0 (X)^2}{(1+(X)^2)(9+(X)^2)} \quad \text{(Eq. 7)}$$

where $X = z/z_0$

$$\Delta\phi = 2\pi I_0(1 - \exp(-L\alpha_0))/\lambda\alpha_0 \quad \text{(Eq. 8)}$$

The nonlinear susceptibility $\chi(3)$ can be obtained by Equation 9, where ε_0 is the vacuum permittivity, c is the speed of light, and n_0 is the refractive index of the medium. The calculated results are shown in Table 3.

$$|\chi^{(3)}| = \sqrt{|cn_0^2/(80\pi)n_2| + |((9\times10^8\varepsilon_0 n_0^2 c^2)/4\pi\omega)\beta|^2} \quad \text{(Eq. 9)}$$

where $\omega = 2\pi c/\lambda$

The results indicated both nonlinear absorption and nonlinear refraction were observed with aperture and without aperture (Figure 4) (Wang et al. 2008b). Further, the conjugative linkage of oxadiazole moiety to the polyacetylene may result in better optical limiting properties, as observed Wang and coworkers (Wang et al. 2008b) (Figure 5). Better optical limiting properties were witnessed with PO3 compared to PO4 (Table 4), which was attributed to better delocalization of electrons between the polymer backbone and the oxadiazole moieties.

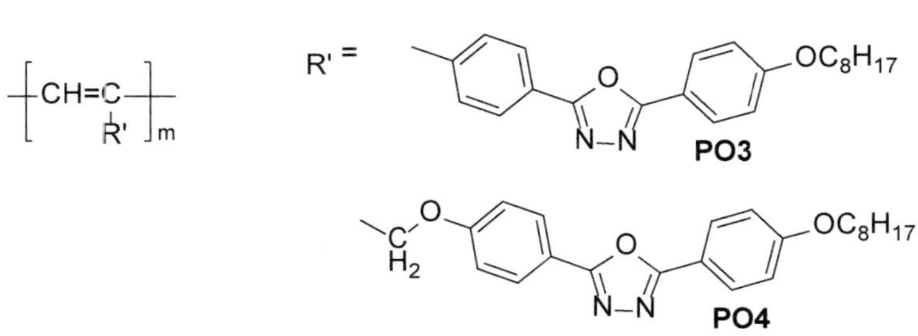

Structures PO3 and PO4

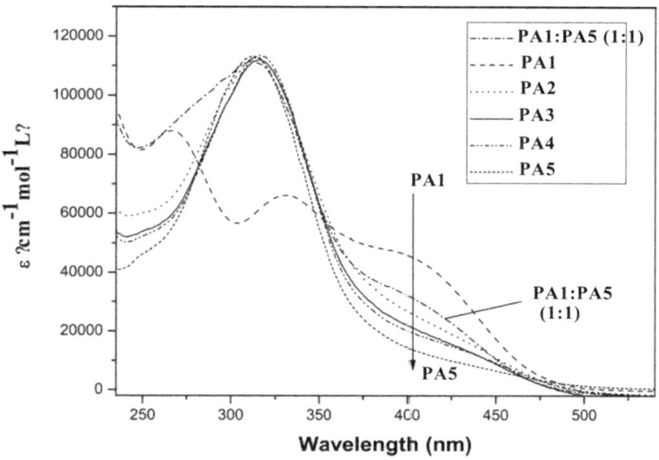

Figure 5. UV–vis spectra of polymers PA1–PA5 and a mixture of PA1:PA5 (1:1) in THF solution. Reproduced with the permission of John Wiley and Sons (Wang et al. 2010).

Table 4. Optical limiting properties of PO3 and PO4. Reproduced here with the permission of RSC John Wiley and Sons (Wang et al. 2008b).

Polymer	Optical Limiting Amplitude (J/cm^2)[a]	Nonlinear Optical Properties[b]		
		α_2 (m/W)	n_2 (m^2/W)	$\chi^{(3)}$ (esu)
PO3	0.305	9.50×10^{-11}	8.98×10^{-19}	3.69×10^{-12}
PO4	0.671	5.60×10^{-11}	7.52×10^{-19}	2.51×10^{-12}

[a] Maximum output intensity.
[b] Measured by the Z-scan technique with an 13-ns Nd:YAG laser system at a 1-Hz repetition rate and a 532-nm wavelength.

2.2 Polyacetylenes Containing Oxadiazole and Thiophene Pendant Groups

Another heterocyclic moiety like thiophene was envisaged to improve the optical limiting properties of polyacetylene polymers with oxadiazole segments. Thiophene unit is also well known for its second-order nonlinear optical properties. However, thiophene-functionalized polyacetylene shows lower thermal stabilities and poor solubilities in common organic solvents (Percec et al. 2002; Wang et al. 2008b; Hegde et al. 2009). Therefore, a combination of thiophene and oxadiazole was a substitute on the polyacetylene chain that was reported to have higher thermal stability and good solubility characteristics. In expectation of improvement in various properties (Wang et al. 2008b), have reported various random polymeric materials based on polyacetylene having pendant oxadiazole and thiophene units (Scheme 2, PA1–PA5). It was observed that the copolymers (PA2–PA5) displayed solubility in common organic solvents (Wang et al. 2010).

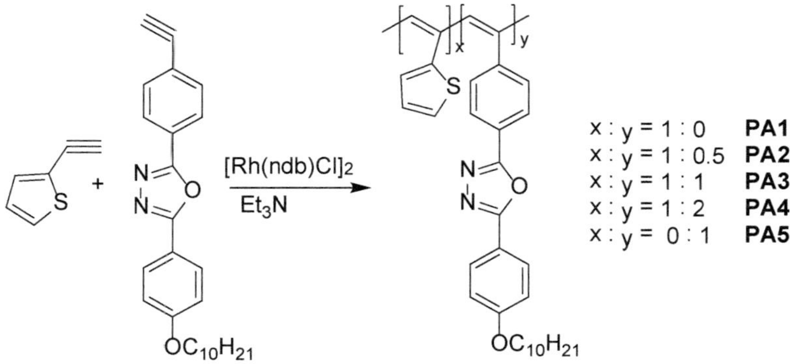

Scheme 2. Synthetic scheme depicting the synthesis of a polyacetylene substituted with thiophene and oxadiazole units.

An improvement in thermal stability was witnessed in incorporating the diaryl-oxadiazole segment into polyacetylene having thiophene units, perhaps due to the jacket effect of diaryl-oxadiazole units. Moreover, the effect of thiophene units hindering or removing a few 6π electrocyclization steps of oxadiazole-containing polyacetylene also aids the thermal stability. The presence of more regular alternating thiophene and oxadiazole pendants along the polymer backbone, like in PA3, further improves the thermal stability. The thermal properties of various polymers were witnessed using thermogravimetric analysis. The polymer PA1 at an M_w of 77900 (M_w/M_n = 1.92) displayed a Td (Thermal decomposition temperature) of 296°C. The polymer PA2 was polymerized to obtain an M_w of 88,700 (M_w/M_n = 2.64) and displayed a Td of 356°C. The polymer PA3 was polymerized to achieve a M_w of 83,600 (M_w/M_n = 2.60) and displayed a Td value of 388°C. The polymer PA4 was polymerized to achieve an M_w of 71,400 (M_w/M_n = 2.98) and displayed a Td value of 378°C, while PA5 was polymerized to achieve an M_w of 25,400 (M_w/M_n = 1.71) and displayed a Td of 353°C. The UV-vis spectra of polymers PA1-PA5 and a 1:1 mixture of PA1 and PA5 were measured in THF. For homopolymer PA1, absorption bands at 268, 332, and 406 nm were observed, while polymers PA2-PA5 displayed an absorption band near 315 nm, which was ascribed to π to π^* transition of oxadiazole fragment. The absorption due to the backbone of PA2-PA5 was observed above 400 nm with low intensities compared to absorptions due to the PA1 backbone. The presence of bulky diaryl-oxadiazole segment on the polyacetylene chain led to a loss of planarity, which reduces the conjugation and, as a result, reduces the intensity of the absorption band, which can also be confirmed by comparing intensities of the absorption band of polymers PA1 to PA5 having increasing content of oxadiazole fragment (Xu et al. 2006; Wang et al. 2010). The observation was also supported by comparing UV-vis spectra of polymers PA1 to PA5 with a mixture (PA1:PA5, 1:1) (Figure 5), which suggested the linkage of a different fragment to the polymer backbone.

The UV-vis spectra shown in Figure 5 also indicate that all polymers display a low intensity at 532 nm, which reduces the loss of intensity with small temperature changes. The photoluminescence spectra recorded for polymers PA1 to PA5 are shown in Figure 6. For polymer PA1, a weak emission was observed, while for polymer PA2–PA5, an enhanced emission was observed due to the presence of diaryl-oxadiazole segments despite having similar emission profiles, which suggested a similar origin of emission having contribution from polymer backbone and diaryl-oxadiazole group. The presence of the thiophene group reduced the stacking interaction due to diaryl-oxadiazole appended to the polymer backbone, which was confirmed due to a blue shift of 7 nm observed in homopolymers PA5 as compared to PA1 caused by disruption of interchain excimer pairs of diaryl-oxadiazole pendants (Wang et al. 2010).

Nonlinear optical property measurements were reported using the Z-scan method and are shown in Table 5. The results suggested that at a feed ratio of greater than 1, the value of χ(3) of PA4 was less than the PA3 and PA2, which was due to the greater content of diaryl-oxadiazole segments twisting the molecular place of the polymer backbone and decreasing the electronic delocalization.

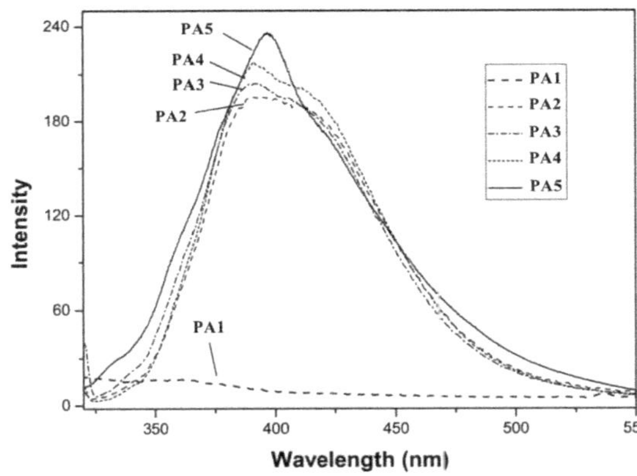

Figure 6. PL spectra for PA1–PA5 in THF solution (2×10^{-5} g/mL). Reproduced here with the permission of John Wiley and Sons (Wang et al. 2010).

Table 5. Summary of Optical Properties of PA1–PA5. Reproduced with the permission of John Wiley and Sons (Wang et al. 2010).

	λ_{abs}[a] (nm)	λ_{em}[a] (nm)	β^b ($\times 10^{-10}$ m/W)	n_2^b ($\times 10^{-17}$ m^2/W)	$\chi^{(3)b}$ ($\times 10^{-11}$ esu)
PA1	266	–	10.9	14.5	10.9
PA2	313	391	9.53	13.5	10.6
PA3	314	391	11.4	14.7	11.0
PA4	317	390	6.59	1.42	1.07
PA5	315	398	0.95	0.09	0.369

[a] Measured in THF solution.
[b] Measured by *Z*-scan technique with 4 ns Nd:YAG laser system at 1 Hz repetition rate and 532 nm wavelength.

2.3 *Poly(3,4-Dialkoxythiophene)s-1,3,4-Oxadiazole Derivative Conjugates*

Owing to the potential electro-optic effect associated with conjugated polymers and NLO response, these are often used in many applications like data storage, optical switching, modulation, optical computing, optical limiting, and so on (Perry et al. 1996; Hegde et al. 2011; Yesodha et al. 2004). Among conjugated polymers having excellent third-order nonlinearities as a result of electron delocalization in their mainchain, which impart them high molecular polarizability, thiophene-based polymers are well suited due to easy processibility, film forming ability, chemical stability, optical transparency, ease of functionalization and mechanical strength (Nisoli et al. 1993; Hegde et al. 2011; Moroni et al. 1994; Kishino et al. 1998). The NLO properties of polythiophenes can also be tuned by incorporating electron donor or acceptor units in the polymer backbone, improving electron delocalization, which enhances the NLO behavior (Hegde et al. 2011). Considering this fact, Hegde et al. have incorporated oxadiazole derivatives in the backbone of polymeric materials (PH1 and PH2) (Hegde et al. 2011).

Incorporating biphenyl segments between 3,4-dialkoxy-substituted thiophen-2-yloxadiazole systems improved electron acceptance to the polymer chain. The polymer was designed to reduce repulsion between the bulky alkoxythiophene groups and increase the planarity to enhance the delocalization of electrons. Analysis of the polymer using thermogravimetric analysis (TGA) revealed its stability up to 310°C. Cyclic voltammetric studies provided a 2.16 eV HOMO-LUMO gap for polymer PH1 and 2.22 eV for PH2.

The UV-vis and fluorescence spectra of PH1 and PH2 were recorded in DMF absorption bands at 368 nm and 376 nm (Figure 7), respectively. For PH1, the emission was witnessed at 392 nm, while for PH2, the emission band was observed at 398 nm in a diluted DMF solution (Figure 7) using an excitation wavelength of 368 nm. The optical properties of PH1 and PH2 suggested that the length of the alkoxy group does not affect the UV-vis or fluorescence spectra. Moreover, wavelength 532 nm is close to the absorption band in the polymers PH1 and PH2. The linear absorption coefficients (α) of the polymers are shown in Figure 8. Table 6 provides the linear and nonlinear optical properties parameters for the polymers. The strong nonlinear property exhibited by both polymers is visible in Figure 7, which fits well with a 3-photon absorption type process described by Equation 10.

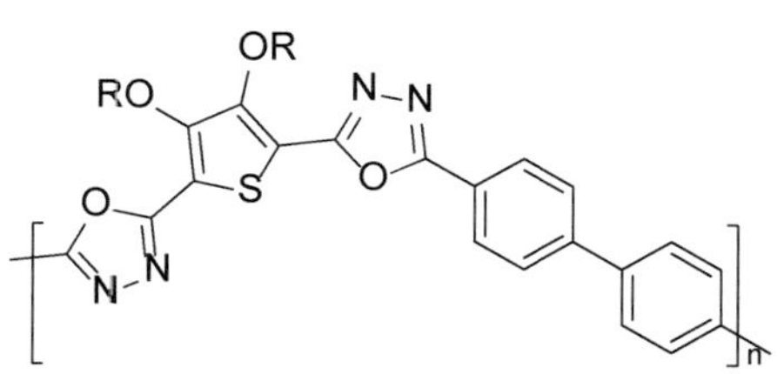

PH1, R = C_7H_{15}
PH2, R = $C_{14}H_{29}$

Structure PH1 and PH2

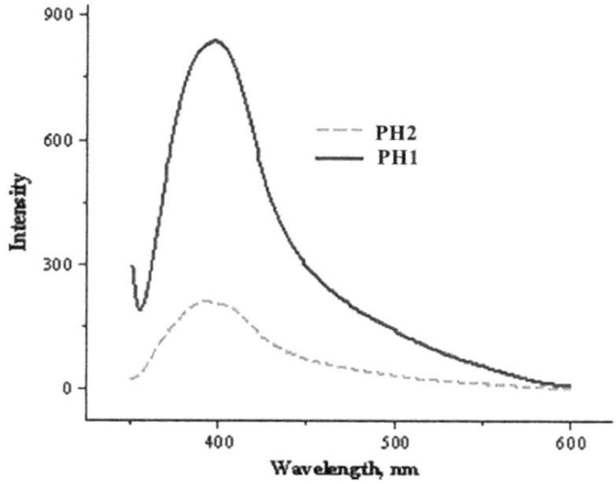

Figure 7. Fluorescence emission spectra of the polymers in DMF solution. Reproduced here with the permission of John Wiley and Sons (Hegde et al. 2011).

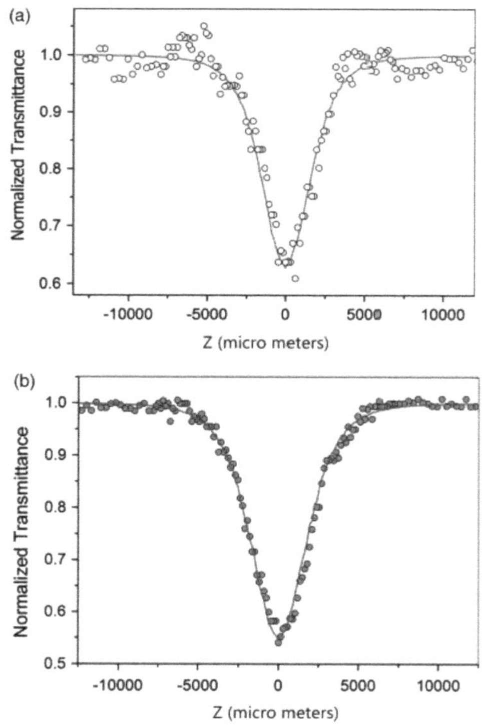

Figure 8. Open-aperture Z-scan curves for (a) PH1 and (b) PH2. Circles are data points, and solid curves are numerical fits obtained using Equation (10). Reproduced here with the permission of John Wiley and Sons (Hegde et al. 2011).

Table 6. Linear and nonlinear optical parameters for the polymers. Reproduced here with the permission of John Wiley and Sons (Hegde et al. 2011).

	Linear optical properties		Nonlinear optical properties (Z-scan)
Sample	n_0[a]	α[b] (m^{-1})	γ ($\times 10^{-24}$ m^3 W^{-2})
PH1	1.432	510	9.0
PH2	1.433	494	17

[a] Refractive index.
[b] Absorption coefficient.

$$T = \frac{(1-R)^2 \exp(-\alpha L)}{\sqrt{\pi \rho_0}} \int_{-\infty}^{+\infty} \ln[\sqrt{1} + \rho_0^2 \exp(-2t^2) + \rho_0 \exp(-t^2)]dt \quad \text{(Eq. 10)}$$

The term "T" represents the transmission of the sample, "R" represents the Fresnel reflection coefficient at the sample-air interface, and the sample length is represented by "L." The term "ρ0" can be expressed as $[2\gamma(1-R)^2 I_0^2 L_{eff}]^{1/2}$, where "γ" represents the 3PA coefficient. The term "L_{eff}" is expressed as $[1-\exp(-2\alpha L)]/2\alpha$.

The study further reveals the length of the alkoxy chain does not affect the nonlinear property of the conjugated polymer-bearing oxadiazole segments.

Sunitha et al. have investigated the NLO properties of conjugated polymers PH3 and PH4 consisting of 1,3,4-oxadiazole, and two thiophene derivatives (3,4-dibenzyloxythiphene, and 3,4-alkoxythiphene) (Sunitha et al. 2012).

Cyclic voltammetric studies were employed to measure the redox potential, which was used to obtain HOMO and LUMO orbital energies and band gap values (Table 7) (de Leeuw et al. 1997). The reduction potential of the polymers PH3 and PH4 was observed to be lower than a widely used electron transport material 2-(4-tert-butylphenyl)-1,3,4-oxadiazole (Table 7) (de Leeuw et al. 1997; Janietz et al. 1997). The polymer PH4 exhibited a higher band gap value than PH3, which was ascribed to a difference in alkoxy group chain length that affects the electronic energy levels by twisting monomer units out of the plane through steric hindrance. The loss of planarity results in a reduction in conjugation.

UV-visible spectrophotometric measurements of the PH3 and PH4 solutions were also measured and compared. For PH3, an absorption band at 388 nm was witnessed in the solution, while it appeared at 395 nm in the polymer film (Figure 9).

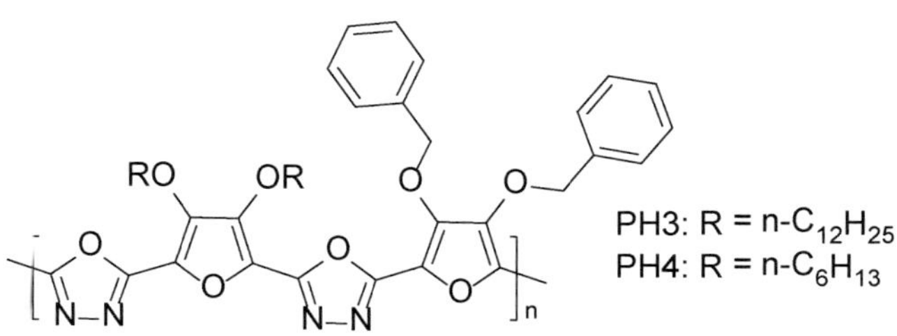

Structures PH3 and PH4

Table 7. Electrochemical potentials, energy levels, and electrochemical band gap of PH3 and PH4. Reproduced here with the permission of the Chemical Society of Japan (Sunitha et al. 2012).

	E_{oxd} (onset)	E_{red} (onset)	E_{HOMO} /eV	E_{LUMO} /eV	E_g^a /eV	E_g^b /eV
PH3	1.37	−1.05	−5.77	−3.35	2.42	2.45
PH4	1.01	−1.19	−5.41	−3.21	2.20	2.38

[a]Electrochemical band gap. [b]Optical band gap.

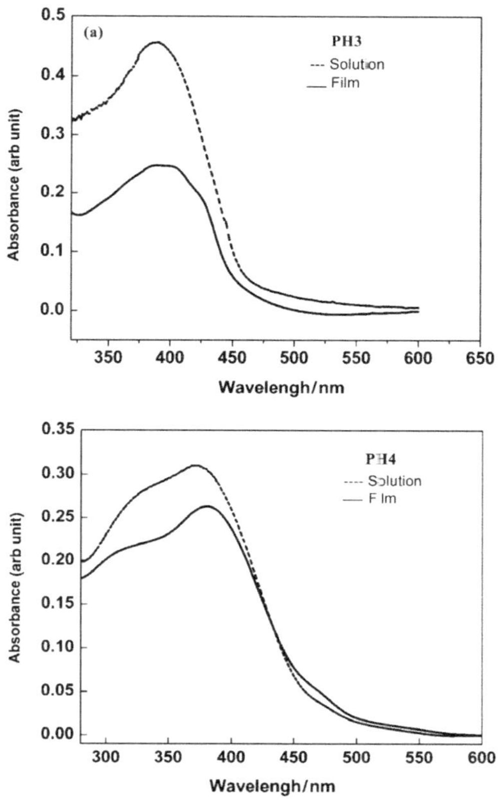

Figure 9. UV-visible absorption spectra of PH3 and PH4 in solution and film states. Reproduced here with the permission of the Chemical Society of Japan (Sunitha et al. 2012).

For PH4, A solution in THF provided an absorption band at 372 nm, while in the case of polymer films, the band was observed at 379 nm. A bathochromic shift of 7 nm in polymer film was witnessed compared to the solution state, suggesting interchain interactions in solid state film of polymers.

For polymers PH3 and PH4, emission bands at 502 and 480 nm were witnessed in THF using an excitation wavelength of 380 nm (Figure 10). Figure 11 and Figure 12 display the open aperture Z-scans and fluence curves of PH3 and PH4, respectively.

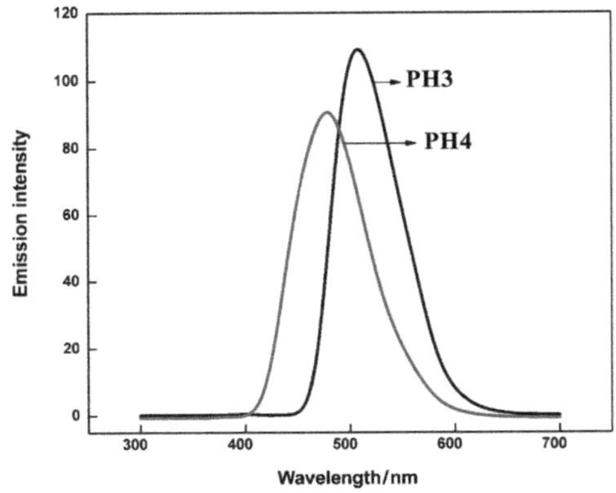

Figure 10. Photoluminescence spectra of PH3 and PH4. Reproduced here with the permission of the Chemical Society of Japan (Sunitha et al. 2012).

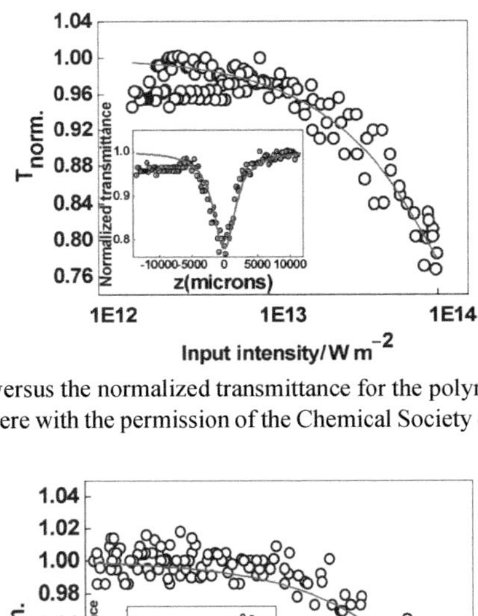

Figure 11. Input intensity versus the normalized transmittance for the polymer PH3. The inset shows the Z-scan curve. Reproduced here with the permission of the Chemical Society of Japan (Sunitha et al. 2012).

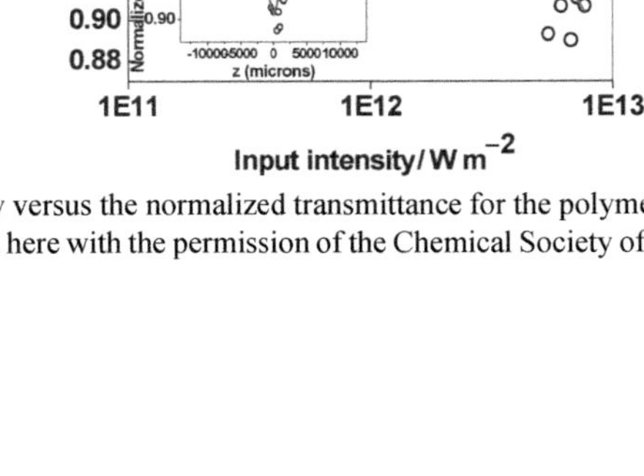

Figure 12. Input intensity versus the normalized transmittance for the polymer PH4. The inset shows the Z-scan curve. Reproduced here with the permission of the Chemical Society of Japan (Sunitha et al. 2012).

A two-photon absorption (TPA) type process was observed to provide the best fit to the Z-scan obtained for both PH3 and PH4 (Equation 11).

$$T(z) = [1/\pi^{1/2}q(z)]\int_{-\infty}^{+\infty} \ln[1+q(z)\exp(-\tau^2)]/d\tau \qquad \text{(Eq. 11)}$$

The term "T(z)" provided sample transmission at the position z, "q(z)" is expressed as $\beta I_0 L/[1+(z/z_0)^2]$, here "$I_0$" represents the peak intensity at the focus point, L = $[1-\exp(-\alpha L)]/\alpha$, "L" is the sample length, "α" represents the linear absorption coefficient. "z_0" is represented by expression $z_0 = \pi\omega_0^2/\lambda$ is the Rayleigh range. The "ω_0" is the beam radium at the focus point, and "β" is the effective two-photon absorption coefficient.

Both polymers provided a linear absorption of about 50% at the excitation wavelength at 1 mm cell length. The experimental data was observed to fit well into the nonlinear transmission equation of a TPA process (Karthikeyan et al. 2008). The unique structure of the polymers was observed to be responsible for the enhanced nonlinear properties. The alkoxy and benzyloxy groups at the 3- and 4-positions of the thiophene ring acted as a donor, while the 1,3,4-oxadiazole ring acted as an acceptor, which acted as a D-A system and provided high π-electron density along the polymer chain. The high π-electron density enhanced the polarizability due to increased electronic delocalization in the polymer chain. The alkoxy and benzyloxy groups also facilitate the solubility of the polymers in common organic solvents like THF. The factors stated above together favored a nonlinear absorption process in these polymers as a whole. Due to the collective results of these factors, the third-order nonlinear susceptibility increases tenfold compared to some other reported D-A polymers (Kiran et al. 2006; Hegde et al. 2009). Overall, both PH3 and PH4 display high optical nonlinear properties (Table 8).

Table 8. Figures of merit calculated using 532 nm light. Reproduced here with the permission of the Chemical Society of Japan (Sunitha et al. 2012).

	$n_2/\text{cm W}^{-1}$	W	T	$I/\text{J ms}^{-1}$
PH3	1.142×10^{-10}	5.0	$\ll 1$	3.25×10^9
PH4	7.31×10^{-11}	4.8	$\ll 1$	4.12×10^9

2.4 Oxadiazole-Bisphenol Conjugates

Aromatic chromophoric segments coupled with aromatic heterocyclic rings are known for their high optical response (Leng et al. 2001; Zadrożna and Kaczorowska 2006), making them suitable candidates for developing NLO materials. Therefore, aromatic compounds like bisphenol A derivatives, which display good photophysical characteristics along with a site for reaction with other suitable heterocyclic moieties, are good candidates for developing NLO materials (Frederiksen et al. 2005; Zadrożna and Kaczorowska 2007). NLO properties are enhanced due to the good donor-acceptor capability of substitutions attached to π-conjugated systems, the presence of highly polarized and noncentrosymmetric structures having donor

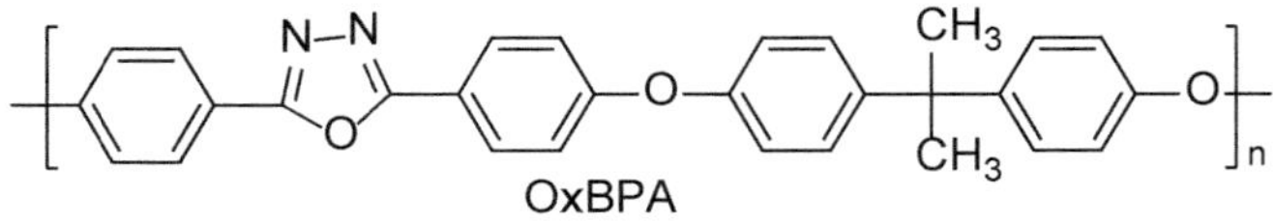

Structure OxBPA

and acceptor substituents, and the existence of intramolecular charge transfer (Li et al. 2006; Morley and Whittaker 2006; Qiu et al. 2013).

As stated above, Homocianu et al. have prepared a polymeric material consisting of bisphenol A and 1,3,4-oxadiazole (OxBPA) and investigated its NLO properties (Homocianu et al. 2019).

Nonlinear properties were investigated using the solvatochromic method. The solvent parameters reported in the study and the estimated polarity parameter (Δf_{LM}) are listed in Table 9. The solvatochromism and solute-solvent interactions for OxBPA were investigated based on equations derived from Lippert-Mataga theory (Equations 12 and 13) (Homocianu et al. 2019).

$$\Delta \nu = \nu^{abs}{}_{\max} - \nu^{em}{}_{\max} = \frac{2(\mu_e - \mu_g)^2 \Delta f_{LM}}{hca^3} + constant \quad \text{(Eq. 12)}$$

$$\Delta f_{LM} = \frac{\varepsilon - 1}{\varepsilon + 2} - \frac{n^2 - 1}{n^2 + 2} \quad \text{(Eq. 13)}$$

The term "Δv" represents the solvatochromic shift (cm^{-1}) between absorption and emission maxima, "ε" represents the relative dielectric constant of the solvent, "n" represents the solvent refractive index, and terms "μ_e" and "μ_g" represent excited

Table 9. Physical parameters of used solvents (ε – the solvent dielectric constant; n – the solvent refractive index; ΔfLM – the polarity parameter) and absorption/fluorescence maxima of the OxBPA sample. The parameters ε and n are reported in the literature (Homocianu et al. 2015).

Media	ε^a	n^b	$\nu_{max}^{abs\ c}$, cm^{-1}	$\nu_{max}^{em\ d}$, cm^{-1}	$\Delta\nu$, [e] cm^{-1}	Δf_{LM} [f]
CHX	2.02	1.4266	33,557	28,011	5546	0.2565
DIO	2.25	1.4224	32,894	28,089	4805	0.2543
TCM	2.24	1.4607	32,786	27,932	4854	0.2742
CLF	4.81	1.4459	32,786	27,777	5009	0.2666
BAc	5.07	1.394	33,112	28,129	4983	0.2392
EtAc	6.08	1.3723	33,444	28,089	5355	0.2274
THF	7.47	1.4070	32,786	28,011	4775	0.2462
DCM	8.93	1.4448	32,679	27,816	4863	0.2660
DCE	10.42	1.4448	32,786	27,816	4970	0.2660
ACN	37.5	1.3442	33,783	27,816	5967	0.2120

CHX-cyclohexane; DIO-dioxane; TCM-tetrachloromethane; CLF-chloroform; BAc-n-butyl acetate; EtAc-ethylacetate; THF-tetrahydrofuran; DCM-dichloromethane; DCE-dichloroethane; ACN-acetonitrile;

[c] Absorption maxima (in cm^{-1}).

[d] Emission maxima (in cm^{-1}).

[e] $\Delta\nu = \nu_{max}^{abs} - \nu_{max}^{em}$ - the Stokes shifts (in cm^{-1}).

[f] Δf_{LM} – the calculated Lippert-Mataga polarity parameter.

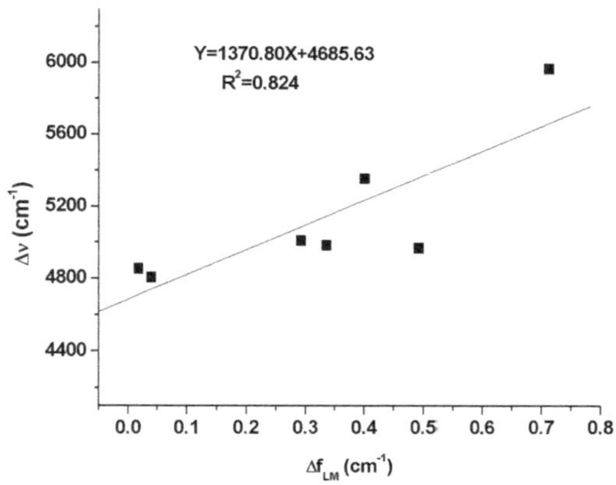

Figure 13. Plots of dependence of Stokes shift (Δv) versus Lippert-Mataga solvent polarity function (ΔfLM) for OxBPA sample (cyclohexane, THF, and DCM solvents were excluded). A solid line is the fitting result with the Lippert-Mataga equation. Reproduced here with the permission of Elsevier (Homocianu et al. 2019).

and ground state dipole moment and the expression $(\mu_e - \mu_g)^2$ is proportional to the slope obtained from Lippert-Mataga plot, "h" is plank constant, "c" is the speed of light in vacuum and a represent Onsager radium of the cavity having solute, which can be calculated from the expression $(3M/4\pi\rho N_A)^{1/3}$ (Suppan 1983; Paley et al. 1989), where N_A is the Avogadro number, while ρ represent the density of the sample calculated using Bicerano method (Bicerano 1996). The Lippert-Mataga Polarity function for OxBPA is shown in Figure 13, which shows that solvatochromic shift varies linearly with Lippert-Mataga solvent polarity, which suggests that polarity causes a spectral change. So, the plot between Stokes shifts and the Lippert-Mataga Polarity function was employed to obtain $\Delta\mu_{CT}$ (dipole moment change) between the excited state and the ground state, and the values obtained are listed in Table 10.

In the case of conjugated systems, the linear optical coefficient of polarizability can be calculated by the expression shown by Equation 14 (Paley et al. 1989; Abbotto et al. 2003).

$$\alpha_{CT} = \alpha_{xx} = 2\frac{\mu_{eg}^{2}}{E_{eg}} = 2\frac{\mu_{eg}^{2}\lambda_{eg}}{hc} = 2\frac{\mu_{eg}^{2}}{hcv} \qquad \text{(Eq. 14)}$$

Where "x" represents the direction of charge transfer and "v" represents absorption frequency. In contrast, "μeg" represents the transition dipole moment, which may be assumed to be the maximum of the bathochromic absorption band and can be calculated using Equation 15 (Carlotti et al. 2012).

$$\mu_{eg}^{2} = \frac{3e^2 h}{8\pi^2 mc} \times \frac{f}{v_{eg}} = 2.13 \times 10^{-30} \times \frac{f}{v_{eg}} \qquad \text{(Eq. 15)}$$

Where "*f*" is the oscillator strength, "h" represents the Plank's constant, and "e" represents the electronic charge. The oscillator strength (*f*) is related to the effective number of electrons involved in the transition from the ground state (S0) to the

Table 10. The nonlinear optical parameters (αCT, βCT, and γCT) and the intramolecular charge transfer characteristics by Mulliken Hush analysis (Cb2, HDA, and RDA) of OxBPA molecules in different solvents. Reproduced here with the permission of Elsevier (Homocianu et al. 2019).

Media	f [a] e.s.u.	μ_{eg} [b] 10^{-35} e.s.u.	$\Delta\mu_{CT}$ [c] 10^{-17} e.s.u.	α_{CT} [d] 10^{-23} e.s.u.	β_{CT} [e] 10^{-29} e.s.u.	γ_{CT} [f] 10^{-34} e.s.u.	C_b^2 [g]	H_{DA} [h] cm^{-1}	R_{DA} [i] Å
CHX	0.2886	1.83	4.85	0.549	2.99	0.13	0.101	10,106	9.37
DIO	0.1517	0.98	4.85	0.300	1.67	7.90	0.036	11,507	5.91
TCM	0.1949	1.27	4.85	0.389	2.16	0.10	0.057	10,958	7.02
CLF	0.1621	1.05	4.85	0.323	1.80	8.52	0.042	11,352	6.18
BAc	0.1666	1.07	4.85	0.326	1.79	8.42	0.043	11,430	6.25
EtAc	0.1677	1.07	4.85	0.321	1.75	8.17	0.043	11,545	6.24
THF	0.1544	1.00	4.85	0.308	1.71	8.13	0.038	11,436	5.98
DCM	0.1523	0.98	4.85	3.005	1.71	8.05	0.037	11,432	5.94
DCE	0.1770	1.14	4.85	0.353	1.96	9.21	0.048	11,535	6.45
ACN	0.1624	1.02	4.85	0.305	1.65	7.57	0.040	11,671	6.08

[a] Oscillator strength.
[b] Transition dipole moment.
[c] Differences between ground and excited state dipole moments determined by using Lippert- Mataga solvent polarity parameter.
[d] The polarizability.
[e] The first hyperpolarizability.
[f] The second hyperpolarizability.
[g] Degree of electronic delocalization.
[h] Electronic coupling matrix (strength of electronic coupling between the ground (S_0) and the charge transfer excited states).
[i] Donor acceptor separation.

excited state (S1) to yield the absorption area in the spectrum. Equation 16 can be used for the calculation of oscillator strength.

$$f = \frac{2.303 m_e c^2}{N_{av} e^2} \int \varepsilon(v) dv = 4.39 \times 10^{-9} \int \varepsilon(v) dv \qquad \text{(Eq. 16)}$$

The term "m_e" represents the mass of the electron in grams, "N_{av}" is the Avogadro number, and integral ∫ε(ν)dν provides the area of the transition band under consideration, which can be calculated using the expression 1.06.ε.$\Delta\nu_{1/2}$ where "ε" represent molar absorption coefficient at the absorption maxima and "$\Delta\nu_{1/2}$" represent the width of the absorption band at $1/2\varepsilon_{max}$. The parameters calculated based on Equations 14–16 are listed in Table 10.

Reproduced here with the permission of Elsevier (Homocianu et al. 2019).

The solvent-dependent first hyperpolarizability can be calculated using Equation 17.

$$\beta_{CT} = \frac{3 v_{eg}^2 \mu_{eg}^2 \Delta\mu_{CT}}{2h^2 c^2 (v_{eg}^2 - v_L^2)(v_{eg}^2 - 4v_L^2)} \qquad \text{(Eq. 17)}$$

Where ν_{eg} is the frequency and assuming $\nu_L = 0$ (at no excitation) expression given in Equation 17 can be written as Equation 18.

$$\beta_{CT} = \beta_{xx} = \frac{\mu_{eg}^{2}\Delta\mu_{CT}}{2(E_{max})^{2}} \quad \text{(Eq. 18)}$$

The calculated values of the first hyperpolarizability β_{CT} are given in Table 10.

The second hyperpolarizability can be calculated using Equation 19 from solvatochromic data.

$$(\gamma_{CT}) = \frac{1}{E_{eg}^{3}}\mu_{eg}^{2}(\Delta\mu^{2} - \Delta\mu_{eg}^{2}) \quad \text{(Eq. 19)}$$

The first hyperpolarizability values of OxBPA calculated using solvatochromic parameters are listed in Table 10, which suggested higher values in cyclohexane than in tetrachloromethane. Moreover, an increase in β_{CT} values with an increase in μ_{eg} was witnessed (Table 10). The C_b^2, H_{DA}, and R_{DA} values calculated using Equations 20–22 are listed in Table 10.

$$C_b^{2} = \frac{1}{2}\left(1 - \sqrt{\frac{\Delta\mu_{eg}^{2}}{\Delta\mu_{eg}^{2} + 4\mu_{eg}^{2}}}\right) \quad \text{(Eq. 20)}$$

$$H_{DA} = \frac{\mu_{eg}^{2}\Delta E_{eg}}{\Delta\mu_{ab}} = \frac{\mu_{eg}^{2}\Delta E_{eg}}{\sqrt{\Delta\mu_{eg}^{2} + 4\mu_{eg}^{2}}} \quad \text{(Eq. 21)}$$

$$R_{DA} = 2.06\times10^{-2}\frac{\Delta E_{eg}\varepsilon_{max}\Delta v_{1/2}}{H_{DA}} \quad \text{(Eq. 22)}$$

A zero value of C_b^2 suggests total delocalization of charges, while a value of one suggests total localization of charges (Erande et al. 2018). A near-zero value of C_b^2 for OxBPA (Table 9) presents a high extent of delocalization and suggests intramolecular charge transfer in all solvents. The parameter H_{DA} indicates the strength of coupling between the ground electronic state (S0) and a charge transfer state, which dictates the charge transfer rate in a D-π-A system and varies inversely with R_{DA} (Donor-acceptor coupling distance). In cyclohexane solution, the lowest H_{DA} and highest R_{DA} values were observed for OxBPA. A slight increase in H_{DA} and R_{DA} values was observed, starting from nonpolar to polar solvent, suggesting intramolecular charge transfer in a polar environment.

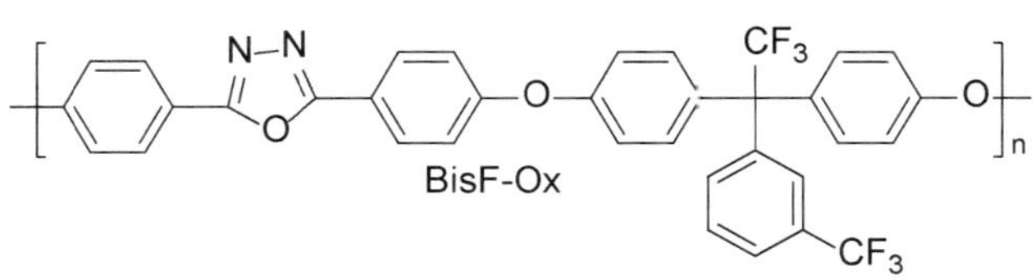

Structure BisF-Ox

Table 11. Nonlinear Optical Properties (NLO) Using the Solvatochromic-Based Approach for BisF-Ox (Homocianu et al. 2022).

Media	f [a] e.s.u.	μ_{eg} [b] 10^{-35} e.s.u.	$\Delta\mu_{CT}$ [c] 10^{-17} e.s.u.	α_{CT} [d] 10^{-23} e.s.u.	β_{CT} [e] 10^{-32} e.s.u.	γ_{CT} [f] 10^{-34} e.s.u.	C_b^2 [g]	H_{DA} [h] cm^{-1}	R_{DA} [i] Å
DIO	0.6671	4.26	6.54	0.128	9.525	0.56	0.196	13,220	3.50
TOL	0.4339	2.79	6.54	0.084	6.320	0.39	0.120	10,745	3.46
$CHCl_3$	0.5527	3.56	6.54	0.108	8.130	0.49	0.162	12,152	3.45
THF	0.6173	3.94	6.54	0.119	8.814	0.53	0.180	12,824	3.47
CH_3OH	0.0495	0.30	6.54	0.086	6.05	0.03	0.001	1602	8.08
DMF	0.5929	3.78	6.54	0.114	8.465	0.50	0.173	12,615	3.46
DMSO	0.6983	4.46	6.54	0.134	0.010	0.58	0.204	13,440	3.52

[a–i]—parameters defined by Equations (1)–(13). **Solvent abbreviations:** DIO (dioxane, $\varepsilon = 2.25$); TOL (toluene, $\varepsilon = 2.38$); $CHCl_3$ (chloroform, $\varepsilon = 4.81$); THF (tetrahydrofuran, $\varepsilon = 7.47$); CH_3OH (methanol, $\varepsilon = 33.00$); DMF (*N,N*-dimethylformamide, $\varepsilon = 38.25$); DMSO (dimethyl sulfoxide, $\varepsilon = 47.24$).

Homocianu et al. recently prepared a fluorinated OxBPA (BisF-Ox) derivative to study their NLO properties (Homocianu et al. 2022). First-order polarizability and second-order hyperpolarizability were estimated using the solvatochromic method, and intramolecular charge transfer properties were evaluated using the Mulliken-Hush approach and listed in Table 11. It was again observed that the NLO properties are affected by the environment around the molecules.

3. Conclusion

In conclusion, various oxadiazole derivatives, including those containing conjugated polymers having main oxadiazole components or covalently attached pendant groups, have been reviewed in this chapter. It was observed that the oxadiazole derivatives having electron donating groups like p-methoxyphenyl or p-dimethylaminophenol donors and electron-deficient acceptors show large second-order linearity. Oxadiazole core can be used to design new NLO molecules where oxadiazole can act as a p-bridge. Solvatochromic parameters can also be used to calculate first-order and second-order hyperpolarizabilities. It is anticipated that the examples listed in the chapter may help design new NLO materials.

References

Abbotto, A., L. Beverina, S. Bradamante, A. Facchetti, C. Klein, G. A. Pagani et al. 2003. A distinctive example of the cooperative interplay of structure and environment in tuning of intramolecular charge transfer in second-order nonlinear optical chromophores. Chem. Eur. J. 9(9): 1991–2007.

Baxamusa, S. 2015. Conformal Polymer CVD. CVD Polymers. K. K. Gleason. Singapore, John Wiley and Sons: 87–109.

Bicerano, J. 1996. Prediction of the properties of polymers from their structures. J. Macromol. Sci. Part C 36(1): 161–196.

Blanchard-Desce, M., C. Runser, A. Fort, M. Barzoukas, J.-M. Lehn, V. Bloy et al. 1995. Large quadratic hyperpolarizabilities with donor-acceptor polyenes functionalized with strong donors. Comparison with donor-acceptor diphenylpolyenes. Chem. Phys. 199(2): 253–261.

Caricato, A. P., W. Ge and A. D. Stiff-Roberts. 2018. UV- and RIR-MAPLE: Fundamentals and Applications. Advances in the Application of Lasers in Materials Science. P. M. Ossi. Cham, Springer International Publishing: 275–308.

Carlotti, B., R. Flamini, I. Kikaš, U. Mazzucato and A. Spalletti. 2012. Intramolecular charge transfer, solvatochromism and hyperpolarizability of compounds bearing ethenylene or ethynylene bridges. Chem. Phys. 407: 9–19.

Cheng, L. T., W. Tam, S. R. Marder, A. E. Stiegman, G. Rikken and C. W. Spangler. 1991. Experimental investigations of organic molecular nonlinear optical polarizabilities. 2. A study of conjugation dependences. J. Phys. Chem. 95(26): 10643–10652.

Cheng, L. T., W. Tam, S. H. Stevenson, G. R. Meredith, G. Rikken and S. R. Marder. 1991. Experimental investigations of organic molecular nonlinear optical polarizabilities. 1. Methods and results on benzene and stilbene derivatives. J. Phys. Chem. 95(26): 10631–10643.

Chopra, K. L., P. D. Paulson and V. Dutta. 2004. Thin-film solar cells: an overview. Prog. Photovolt: Res. Appl. 12(2-3): 69–92.

Chun, H., I. K. Moon, D. H. Shin and N. Kim. 2001. Preparation of highly efficient polymeric photorefractive composite containing an isophorone-based NLO chromophore. Chem. Mater. 13(9): 2813–2817.

Clays, K. and A. Persoons. 1991. Hyper-Rayleigh scattering in solution. Phys. Rev. Lett. 66(23): 2980–2983.

de Leeuw, D. M., M. M. J. Simenon, A. R. Brown and R. E. F. Einerhand. 1997. Stability of n-type doped conducting polymers and consequences for polymeric microelectronic devices. Synth. Met. 87(1): 53–59.

Dhonnar, S. L., V. A. Adole, R. A. More, N. V. Sadgir, B. S. Jagdale, T. B. Pawar et al. 2022. Synthesis, molecular structure, electronic, spectroscopic, NLO and antimicrobial study of N-benzyl-2-(5-aryl-1,3,4-oxadiazol-2-yl)aniline derivatives. J. Mol. Struct. 1262: 133017.

Duan, Y., C. Ju, G. Yang, E. Fron, E. Coutino-Gonzalez, S. Semin et al. 2016. Aggregation induced enhancement of linear and nonlinear optical emission from a hexaphenylene derivative. Adv. Funct. Mater. 26(48): 8968–8977.

Erande, Y., S. Kothavale, M. C. Sreenath, S. Chitrambalam, I. H. Joe and N. Sekar. 2018. Triphenylamine derived coumarin chalcones and their red emitting OBO difluoride complexes: Synthesis, photophysical and NLO property study. Dyes Pigm. 148: 474–491.

Fann, W. S., S. Benson, J. M. J. Madey, S. Etemad, G. L. Baker and F. Kajzar. 1989. Spectrum of ${\ensuremath{\chi}}^{(3)}$(-3\ensuremath{\omega};\ensuremath{\omega},\ensuremath{\omega},\ensuremath{\omega}) in polyacetylene: An application of free-electron laser in nonlinear optical spectroscopy. Phys. Rev. Lett. 62(13): 1492–1495.

Frederiksen, P. K., S. P. McIlroy, C. B. Nielsen, L. Nikolajsen, E. Skovsen, M. Jørgensen et al. 2005. Two-photon photosensitized production of singlet oxygen in water. J. Am. Chem. Phys. 127(1): 255–269.

Ghezelbash, Z. D., H. Motiei, M. Mahmoody and K. A. Delmaghani. 2019. Synthesis, characterization, and nonlinear optical properties of some new series of S-(5-aryl-1, 3, 4-oxadiazol-2-yl) 2-chloroethanethioate derivatives. Turk. J. Chem. 43(3): 902–910.

He, Y. Xiao, Huang, Lin, K. Y. Mya and Zhang. 2004. Highly efficient luminescent organic clusters with quantum dot-like properties. J. Am. Chem. Soc. 126(25): 7792–7793.

Hegde, P. K., A. V. Adhikari, M. G. Manjunatha, P. Poornesh and G. Umesh. 2009. Third-order nonlinear optical susceptibilities of new copolymers containing alternate 3,4-dialkoxythiophene and (1,3,4-oxadiazolyl)pyridine moieties. Opt. Mater. 31(6): 1000–1006.

Hegde, P. K., A. V. Adhikari, M. G. Manjunatha, C. S. Suchand Sandeep and R. Philip. 2009. Synthesis and nonlinear optical characterization of new poly{2,2′-(3,4-didodecyloxythiophene-2,5-diyl)bis[5-(2-thienyl)-1,3,4-oxadiazole]}. Synth. Met. 159(11): 1099–1105.

Hegde, P. K., A. Vasudeva Adhikari, M. G. Manjunatha, C. S. Suchand Sandeep and R. Philip. 2011. Novel poly(3,4-dialkoxythiophene)s carrying 1,3,4-oxadiazolyl-biphenyl moieties: synthesis and nonlinear optical studies. Polym. Int. 60(1): 112–118.

Homocianu, M., A.-M. Ipate, C. Hamciuc and A. Airinei. 2015. Specific spectral characteristics of some phenylquinoxaline derivatives. 202: 62–67.

Homocianu, M., A. Airinei, C. Hamciuc and A. M. Ipate. 2019. Nonlinear optical properties (NLO) and metal ions sensing responses of a polymer containing 1,3,4-oxadiazole and bisphenol A units. J. Mol. Liq. 281: 141–149.

Homocianu, M., A. Airinei, A. M. Ipate and C. Hamciuc. 2022. Spectroscopic recognition of metal ions and non-linear optical (NLO) properties of some fluorinated poly(1,3,4-oxadiazole-ether)s. Chemosensors. 10. DOI: 10.3390/chemosensors10050183.

Janietz, S., A. Wedel and R. Friedrich. 1997. Electrochemical redox behavior of aromatic poly(1,3,4-oxadiazole)s and sulfur containing polymers and their use in LED's. Synth. Met. 84(1): 381–382.

Jin, S.-H., M.-Y. Kim, J. Y. Kim, K. Lee and Y.-S. Gal. 2004. High-efficiency poly(p-phenylenevinylene)-based copolymers containing an oxadiazole pendant group for light-emitting diodes. J. Am. Chem. Soc. 126(8): 2474–2480.

Joule, J. A. and K. Mills. 2012. Heterocyclic Chemistry at a Glance. United Kingdom, John Wiley & Sons.

Karthikeyan, B., M. Anija, C. S. Suchand Sandeep, T. M. Muhammad Nadeer and R. Philip. 2008. Optical and nonlinear optical properties of copper nanocomposite glasses annealed near the glass softening temperature. Opt. Commun. 281(10): 2933–2937.

Kiran, A. J., D. Udayakumar, K. Chandrasekharan, A. V. Adhikari and H. D. Shashikala. 2006. Z-scan and degenerate four wave mixing studies on newly synthesized copolymers containing alternating substituted thiophene and 1,3,4-oxadiazole units. J. Phys. B: At., Mol. Opt. Phys. 39(18): 3747.

Kishino, S., Y. Ueno, K. Ochiai, M. Rikukawa, K. Sanui, T. Kobayashi et al. 1998. Estimate of the effective conjugation length of polythiophene from its $|{\ensuremath{\chi}}^{(3)}(\ensuremath{\omega};\ ensuremath{\omega},\ensuremath{\omega},\ensuremath{-}\ensuremath{\omega})|$ spectrum at excitonic resonance. Phys. Rev. B 58(20): R13430–R13433.

Krishnan, A., S. K. Pal, P. Nandakumar, A. G. Samuelson and P. K. Das. 2001. Ferrocenyl donor–organic acceptor complexes for second order nonlinear optics. Chem. Phys. 265(3): 313–322.

Kumar, P. C. R., K. Janardhana, K. M. Balakrishna and T. Sheela. 2016. Study of non linear optical properties of a chalcone doped PVA composite material. Procedia. Eng. 141: 83–90.

Lam, J. W. Y., Y. Dong, K. K. L. Cheuk, J. Luo, Z. Xie, H. S. Kwok et al. 2002. Liquid crystalline and light emitting polyacetylenes: synthesis and properties of biphenyl-containing poly(1-alkynes) with different functional bridges and spacer lengths. Macromolecules 35(4): 1229–1240.

Lam, J. W. Y. and B. Z. Tang. 2005. Functional Polyacetylenes. 38(9): 745–754.

Lee, M., H. E. Katz, C. Erben, D. M. Gill, P. Gopalan, J. D. Heber et al. 2002. Broadband modulation of light by using an electro-optic polymer. Science 298(5597): 1401–1403.

Leng, W. N., Y. M. Zhou, Q. H. Xu and J. Z. Liu. 2001. Synthesis of nonlinear optical polyimides containing benzothiazole moiety and their electro-optical and thermal properties. Polymer 42(22): 9253–9259.

Li, H., K. Han, X. Shen, Z. Lu, Z. Huang, W. Zhang et al. 2006. The first hyperpolarizabilities of hemicyanine cationic derivatives studied by finite-field (FF) calculations. J. Mol. Struct.: Theochem 767(1): 113–118.

Liu, J., J. W. Y. Lam and B. Z. Tang. 2009. Acetylenic polymers: syntheses, structures, and functions. Chem. Rev. 109(11): 5799–5867.

Marder, S. R., L.-T. Cheng, B. G. Tiemann, A. C. Friedli, M. Blanchard-Desce, J. W. Perry et al. 1994. Large first hyperpolarizabilities in push-pull polyenes by tuning of the bond length alternation and aromaticity. Science 263(5146): 511–514.

Maria, M. 2018. Synthesis and Nonlinear Optical Studies on Organic Compounds in Laser-Deposited Films. Applied Surface Science. I. Gurrappa. Rijeka, IntechOpen: Ch. 1.

Mashraqui, S. H., R. S. Kenny, S. G. Ghadigaonkar, A. Krishnan, M. Bhattacharya and P. K. Das. 2004. Synthesis and nonlinear optical properties of some donor–acceptor oxadiazoles. Opt. Mater. 27(2): 257–260.

Morley, J. O. and S. D. Whittaker. 2006. Non-linear optical properties of thienylmethylene anilines and benzylidene aminothiophenes. J. Mol. Struct.: Theochem 760(1): 1–13.

Moroni, M., J. Le Moigne and S. Luzzati. 1994. Rigid rod conjugated polymers for nonlinear optics: 1. Characterization and linear optical properties of poly(aryleneethynylene) derivatives. Macromolecules 27(2): 562–571.

Moylan, C. R., R. J. Twieg, V. Y. Lee, S. A. Swanson, K. M. Betterton and R. D. Miller. 1993. Nonlinear optical chromophores with large hyperpolarizabilities and enhanced thermal stabilities. J. Am. Chem. Soc. 115(26): 12599–12600.

Mukherjee, S. and P. Thilagar. 2014. Organic white-light emitting materials. Dyes Pigm. 110: 2–27.

Nanjo, K., S. M. A. Karim, R. Nomura, T. Wada, H. Sasabe and T. Masuda. 1999. Synthesis and properties of poly(1-naphthylacetylene) and poly(9-anthrylacetylene). J. Polym. Sci. A Polym. Chem. 37(3): 277–282.

Nie, W. 1993. Optical Nonlinearity: Phenomena, applications, and materials. Adv. Mater. 5(7-8): 520–545.

Nisoli, M., A. Cybo-Ottone, S. De Silvestri, V. Magni, R. Tubino, C. Botta et al. 1993. Femtosecond transient absorption saturation in poly(alkyl-thiophene-vinylene)s. Phys. Rev. B 47(16): 10881–10884.

Nomura, R., S. M. Abdul Karim, H. Kajii, R. Hidayat, K. Yoshino and T. Masuda. 2000. Metathesis polymerization of 9-(10-hexoxycarbonyl)anthrylacetylene. A Route to a Widely Conjugated Polyacetylene with Excellent Stability and Solubility. Macromolecules 33(12): 4313–4315.

Paley, M. S., J. M. Harris, H. Looser, J. C. Baumert, G. C. Bjorklund, D. Jundt et al. 1989. A solvatochromic method for determining second-order polarizabilities of organic molecules. J. Org. Chem. 54(16): 3774–3778.

Percec, V., M. Obata, J. G. Rudick, B. B. De, M. Glodde, T. K. Bera et al. 2002. Synthesis, structural analysis, and visualization of poly(2-ethynyl-9-substituted carbazole)s and poly(3-ethynyl-9-substituted carbazole)s containing chiral and achiral minidendritic substituents. J. Polym. Sci. A Polym. Chem. 40(20): 3509–3533.

Perry, J. W., K. Mansour, I. Y. S. Lee, X. L. Wu, P. V. Bedworth, C. T. Chen et al. 1996. Organic optical limiter with a strong nonlinear absorptive response. Science 273(5281): 1533–1536.

Qiu, Y.-Q., Z. Li, N.-N. Ma, S.-L. Sun, M.-Y. Zhang and P.-J. Liu. 2013. Third-order nonlinear optical properties of molecules containing aromatic diimides: Effects of the aromatic core size and a redox-switchable modification. J. Mol. Graph. Model. 41: 79–88.

Rao, V. P., Y. M. Cai and A. K. Y. Jen. 1994. Ketene dithioacetal as a π-electron donor in second-order nonlinear optical chromophores. J. Chem. Soc., Chem. Commun. (14): 1689–1690.

Rao, V. P., A. K. Y. Jen, K. Y. Wong and K. J. Drost. 1993. Dramatically enhanced second-order nonlinear optical susceptibilities in tricyanovinylthiophene derivatives. J. Chem. Soc., Chem. Commun. (14): 1118–1120.

Sata, T., R. Nomura, T. Wada, H. Sasabe and T. Masuda. 1998. Polymerization of N-carbazolylacetylene by various transition metal catalysts and polymer properties. J. Polym. Sci. A Polym. Chem. 36(14): 2489–2492.

Schweig, A. 1967. Calculation of static electric higher polarizabilities of closed shell organic π-electron systems using a variation method. Chem. Phys. Lett. 1(5): 195–199.

Sheik-Bahae, M., A. A. Said, T. H. Wei, D. J. Hagan and E. W. V. Stryland. 1990. Sensitive measurement of optical nonlinearities using a single beam. IEEE J. Quantum Electron. 26(4): 760–769.

Shi, Y., W. Lin, D. J. Olson, J. H. Bechtel, H. Zhang, W. H. Steier et al. 2000. Electro-optic polymer modulators with 0.8 V half-wave voltage. Appl. Phys. Lett. 77(1): 1–3.

Su, X., H. Xu, Q. Guo, G. Shi, J. Yang, Y. Song et al. 2008. Stilbene-containing polyactylenes: Molecular design, synthesis, and relationship between molecular structure and NLO properties. J. Polym. Sci. A Polym. Chem. 46(13): 4529–4541.

Sung, H.-H. and H.-C. Lin. 2004. Novel alternating fluorene-based conjugated polymers containing oxadiazole pendants with various terminal groups. Macromolecules 37(21): 7945–7954.

Sunitha, M. S., A. V. Adhikari, K. A. Vishnumurthy, N. Smijesh and R. Philip. 2012. Electrochemical and nonlinear optical studies of new D–A Type π-conjugated polymers carrying 3,4-benzyloxythiophene, oxadiazole, and 3,4-alkoxythiophene systems. Chem. Lett. 41(3): 234–236.

Suppan, P. 1983. Excited-state dipole moments from absorption/fluorescence solvatochromic ratios. Chem. Phys. Lett. 94(3): 272–275.

Tang, B. Z., X. Kong, X. Wan and X.-D. Feng. 1997. Synthesis and properties of stereoregular polyacetylenes containing cyano groups, poly[[4-[[[n-[(4'-cyano-4-biphenylyl)-oxy]alkyl]oxy] carbonyl]phenyl]acetylenes]. Macromolecules 30(19): 5620–5628.

Tathe, A. B. and N. Sekar. 2016. Red emitting NLOphoric 3-styryl coumarins: Experimental and computational studies. Opt. Mater. 51: 121–127.

Wang, J. F., G. E. Jabbour, E. A. Mash, J. Anderson, Y. Zhang, P. A. Lee et al. 1999. Oxadiazole metal complex for organic light-emitting diodes. Adv. Mater. 11(15): 1266–1269.

Wang, P., C. Chai, Q. Yang, F. Wang, Z. Shen, H. Guo et al. 2008. Synthesis and characterization of bipolar copolymers containing oxadiazole and carbazole pendant groups and their application to electroluminescent devices. J. Polym. Sci. A Polym. Chem. 46(16): 5452–5460.

Wang, X., S. Guan, H. Xu, X. Su, X. Zhu and C. Li. 2010. Preparation and properties of oxadiazole-containing polyacetylenes as electron transport materials. J. Polym. Sci. A Polym. Chem. 48(6): 1406–1414.

Wang, X., S. Guang, H. Xu, S. Xinyan, J. Yang, Y. Song et al. 2008a. Thermally stable oxadiazole-containing polyacetylenes: Relationship between molecular structure and nonlinear optical properties. J. Mater. Chem. 18(35): 4204–4209.

Wang, X., J. Wu, H. Xu, P. Wang and B. Z. Tang. 2008b. Preparation and property of two soluble oxadiazole-containing functional polyacetylenes. J. Polym. Sci. A 46(6): 2072–2083.

Wang, X., Y. Yan, T. Liu, X. Su, L. Qian, Y. Song et al. 2010. Synthesis and nonlinear optical properties of polyacetylenes containing oxadiazole and thiophene pendant groups with high thermal stability. J. Polym. Sci. A 48(23): 5498–5504.

Xu, H., B. Yang, X. Gao, C. Li and S. Guang. 2006. Synthesis and characterization of organic–inorganic hybrid polymers with a well-defined structure from diamines and epoxy-functionalized polyhedral oligomeric silsesquioxanes. J. Appl. Polym. Sci. 101(6): 3730–3735.

Yesodha, S. K., C. K. Sadashiva Pillai and N. Tsutsumi. 2004. Stable polymeric materials for nonlinear optics: a review based on azobenzene systems. Prog. Polym. Sci. 29(1): 45–74.

Yin, S., H. Xu, M. Fang, W. Shi, Y. Gao and Y. Song. 2005. Synthesis and optical properties of poly[2-{n-methyl-n-(4-(4-ethynylphenylazo)phenyl)amino}ethyl butyrate]. Macromol. Chem. Phys. 206(15): 1549–1557.

Yin, S., H. Xu, X. Su, L. Wu, Y. Song and B. Z. Tang. 2007. Preparation and property of soluble azobenzene-containing substituted poly(1-alkyne)s optical limiting materials. Dyes Pigm. 75(3): 675–680.

Zadrożna, I. and E. Kaczorowska. 2006. Synthesis and characteristics of azo chromophores for nonlinear-optical application. Dyes Pigm. 71(3): 207–211.

Zadrożna, I. and E. Kaczorowska. 2007. The theoretical and experimental investigations of first hyperpolarizabilities for azo chromophores—Derivatives of bisphenol A. Mater. Chem. Phys. 103(2): 465–469.

Chapter 6

Oxadiazole-Based Heterocyclic Compounds as Chemical Probes in Live Cell Imaging

1. Introduction

Oxadiazoles are five-membered heterocyclic compounds that belong to the aromatic azole family. Out of four different isomers of oxadiazole, 1,2,4-oxadiazole, and 1,3,4-oxadiazole exhibit a diverse range of chemical as well as biological characteristics. These heterocyclic scaffolds (1,3,4-oxadiazole) are used as primary synthons during drug manufacturing (Siwach and Verma 2020). Additionally, 1,3,4-oxadiazole-based probes exhibit high fluorescence quantum yield and good thermal and chemical stability (Hughes and Bryce 2005). These electron-deficient heterocyclic compounds can unravel remarkable biological phenomena when linked with suitable functional moieties. For example, (i) attachment of ^{18}F ligand to the basic oxadiazole derivative could potentially map abnormal tissues in the body, and (ii) modification of oxadiazole-based structure with $^{123/125}I$ could map the exact location of cholinergic nerve loss in case of dementia.

Two of its positions on the ring can be substituted with different groups to yield a range of useful compounds (Maan et al. 2021). The covalent attachment of oxadiazole with fluorophores provides a brighter fluorescence response. These fluorescent probes could be utilized for cell function labeling, pH indicators, and intracellular metal ion quantification. Oxadiazole derivatives are also being explored as chelating motifs owing to their rich coordination characteristics. Potential metal ions such as Zn^{2+} and Fe^{3+}/Fe^{2+} might be suitable for binding with the oxadiazole precursor in the presence of a suitable functional group such as hydroxyl moiety.

2. Oxadiazole-Based Probes in PET Imaging

Positron emission tomography (PET) is one of the advanced non-invasive imaging tools to investigate the biochemical nature of tissues and organs. During the PET scanning process, a radioactive drug called "tracer" is normally used in order to visualize the bright signals. This type of imaging is more suitable for abnormal tissue detection; therefore, it is more frequently used in cancer detection. Applications of oxadiazole-based synthetic probes as radiotracers in PET imaging are explained below.

The subtype sphingosine 1-phosphate receptor 1 (S1PR1) exhibits inflammatory importance among G-protein-coupled receptors. Rosenberg et al. developed fluorine (^{18}F) tagged S1P probe 1 in a multi-step synthetic process (Rosenberg et al. 2016). The synthetic ligand that contains centrally positioned 1,2,4-oxadiazole moiety exhibited high selectivity (> 100-fold) as well as high potency (2.63 nM) toward $S1P_1$ and accumulated in the mouse liver during *in vivo* evaluation. Studies like microPET imaging, *in vitro* autoradiography, and biodistribution confirmed ^{18}F tagged ligand 1 as a positron emission tomography (PET) radiotracer for $S1P_1$ and could be used for LPS-induced liver injury with increased expression of $S1P_1$. Luo et al. developed triazole functionalized probe 2 by modifying the earlier version of azetidine-3-carboxylic acid-based ligand with hydroxymethyl capped triazole moiety with around 14.1% radiochemical yield and 54.1 GBq/μ mol specific activity (Luo et al. 2018). The biodistribution analysis found interesting results where the fluorinated probe can enter the blood-brain barrier with 0.71% ID/g brain uptake following 1 h of injection (Figure 1). Additionally, the uptake was shortened by a specific S1PR1 ligand called SEW2871, confirmed by *in vitro* autoradiography analysis. The ^{18}F labeled probe might be useful for inflammatory evaluation as a PET radiotracer.

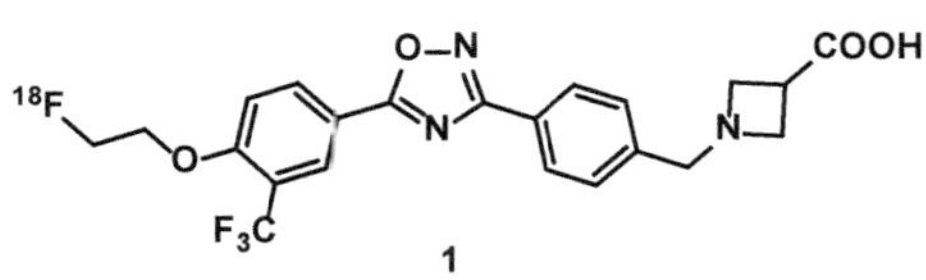
1

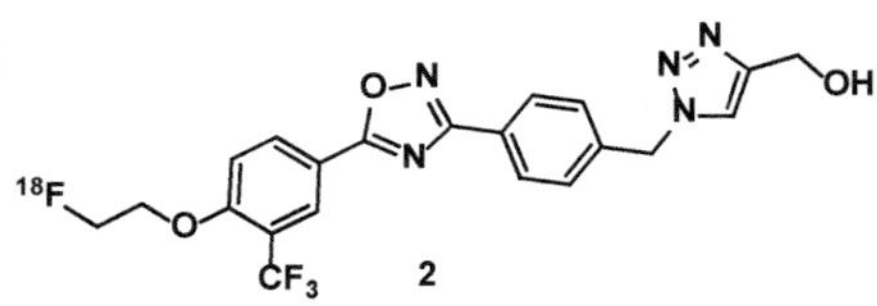
2

Structures 1 and 2

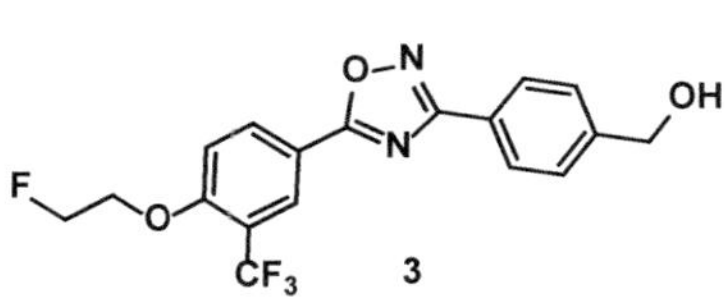
3

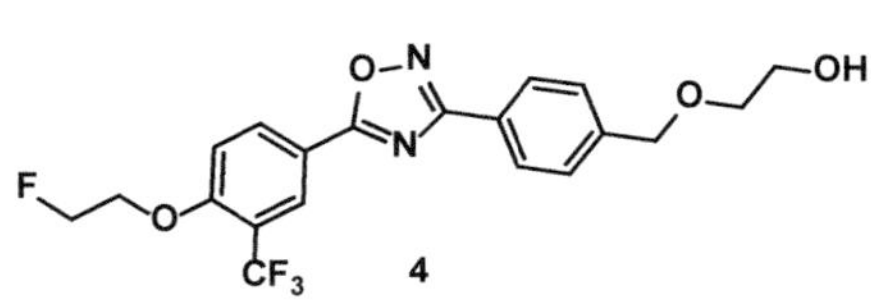
4

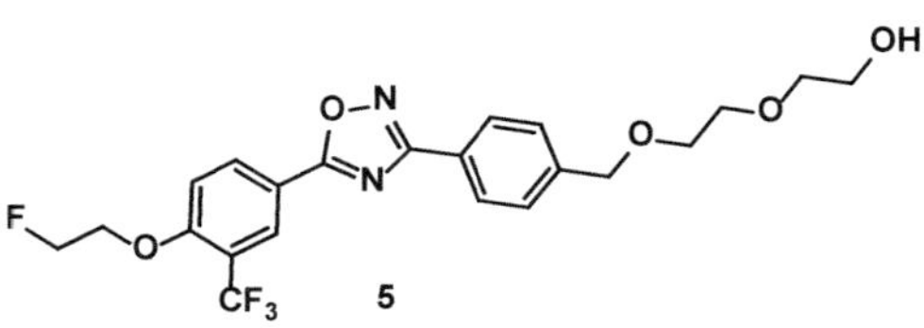
5

Structures 3–5

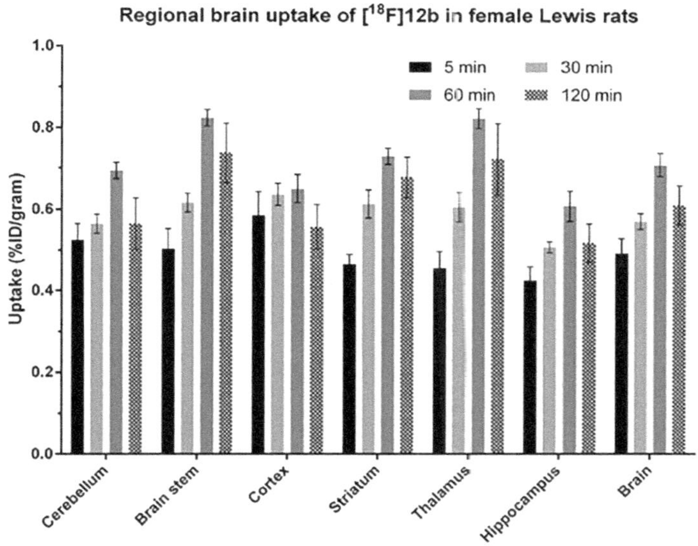

Figure 1. A bar chart showing regional brain uptake distribution of probe 2 in female Lewis rats. Animals were euthanized after 5, 30, 60, and 120 min injections. %ID/gram values (mean ± SD) with 4 rats per group. Reprinted with permission from Elsevier (Luo et al. 2018).

In 2018, The same group (Luo et al.) also developed a series of fluorine-substituted S1PR1 probes 3–5 (Luo et al. 2018). The compounds 3–5 displayed strong binding tendencies toward the S1PR1 ligand with IC_{50} values of 6.7 ± 0.7, 14.0 ± 0.4, and 15.4 ± 3.3 nM, respectively. Radiolabelling these compounds with F-18 resulted in high-yield compounds with more than 98% purity. The synthesized radiolabelled probes could penetrate the blood-brain barrier, which might be useful for *in vivo* response in inflammatory models in the future.

Visualization of type 2 cannabinoid receptor (CB2) has significant advantages in quantification of neuro-inflammation and clinical settings. The CB2 receptor is mainly expressed in immune cells (Gazzi et al. 2022). Ahamed et al. developed N-arylamide functionalized oxadiazole derivatives labeled with ^{11}C and ^{18}F (Ahamed et al. 2016). The binding affinity for the human CB2 receptor was found to be 87 nM and 0.8 nM with probes 6 and 8, respectively. The EC_{50} values for probes 6 and 8 were estimated to be 3 and 0.1 nM, respectively, which were confirmed as potent CB2 receptor agonists. The synthesized probes 7 and 9 exhibited good brain uptake along with appropriate clearance from blood, which suggested that these probes could be useful for PET *in vivo* imaging of CB2 receptors. In another study, probe 9 exhibited a longer half-life (109.8 min) and greater affinity toward CB2 receptor agonist ($K_i \sim 0.8$ nM) (Attili et al. 2019).

Glycogen synthase kinase 3 beta (GSK-3β) is a well-known Ser/Thr protein kinase enzyme that is a tumor suppressor (Mishra 2010). Kumata et al. developed ^{11}C labeled radiotracers 10–12 using thiol-based synthetic precursor (Kumata

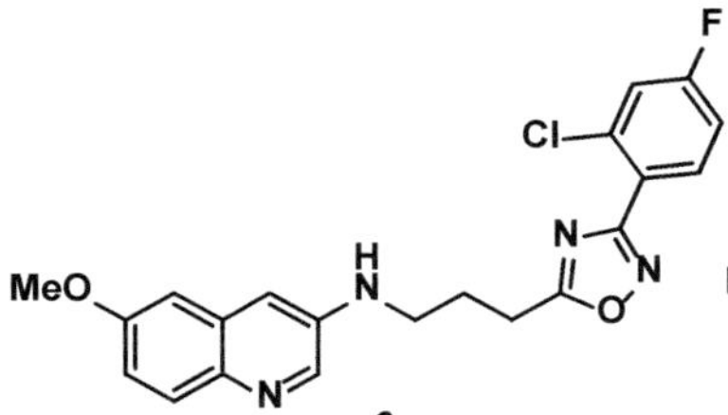

6

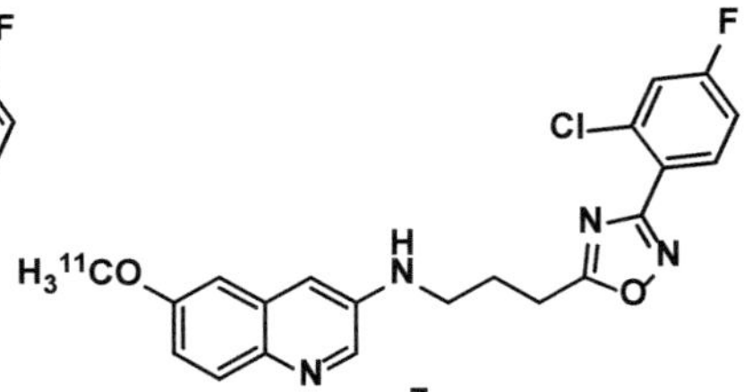

7

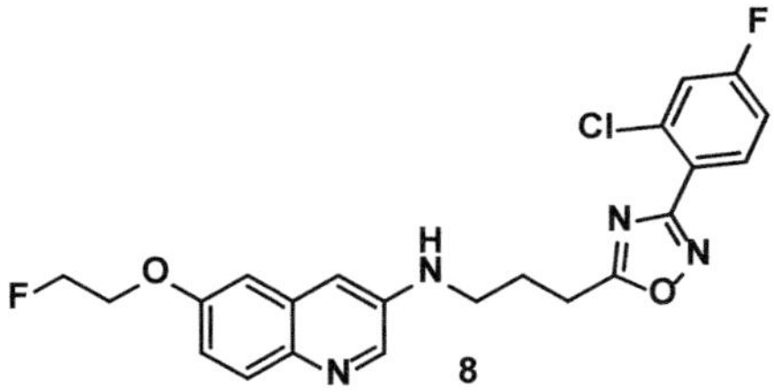

8

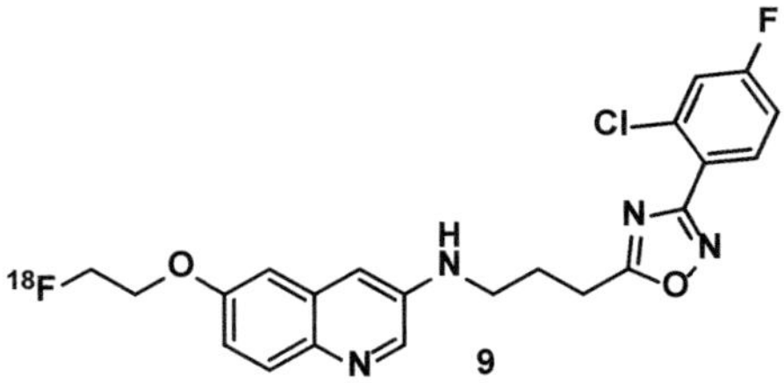

9

Structures 6–9

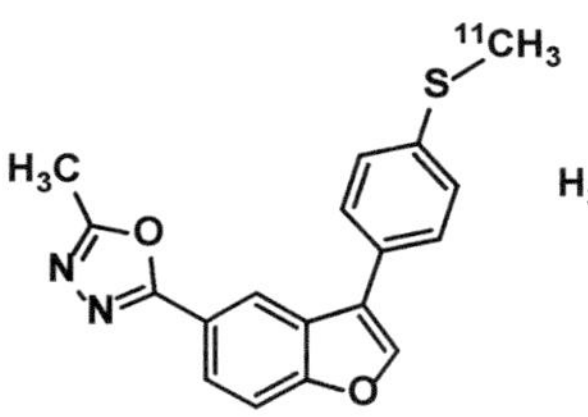

10

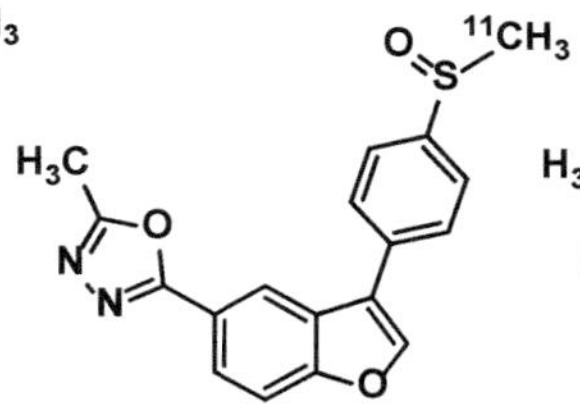

11

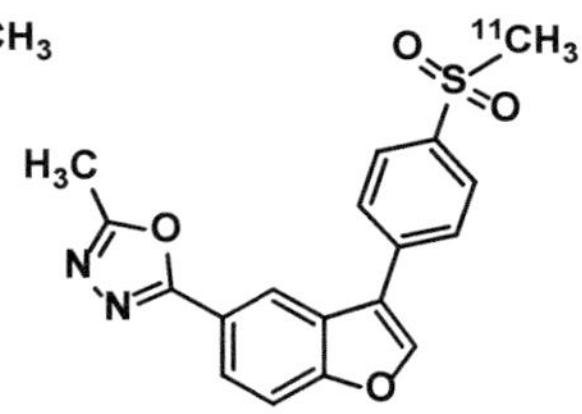

12

Structures 10–12

et al. 2015). Both the methylsulfanyl (10) and methylsulfonyl (12) probes exhibited elevated expression of GSK-3β in CWS (cold water stress) mouse brains as a result of significant uptake of radioactivity. These radiotracers could be used as GSK-3β inhibitors.

The histone deacetylase (HDAC) enzyme eliminates acetyl moiety from the lysine part of histone protein through the deacetylation process. This catalytic method regulates oncogene expression and cell-cycle progression (Ropero and Esteller 2007). HDAC is known as a promising target for chemotherapy. The trifluoromethyloxadiazolyl (TFMO) is a unique class-IIa HDAC pharmacophore that coordinates with the Zn^{2+} ion in the catalytic pocket. Therefore, Turkman et al. developed an ^{18}F labeled oxadiazole scaffold (> 98% radiochemical purity) through radiofluorination from bromodifluoromethyl attached oxadiazole (Turkman et al. 2021). The probe worked as a class-IIa HDAC inhibitor.

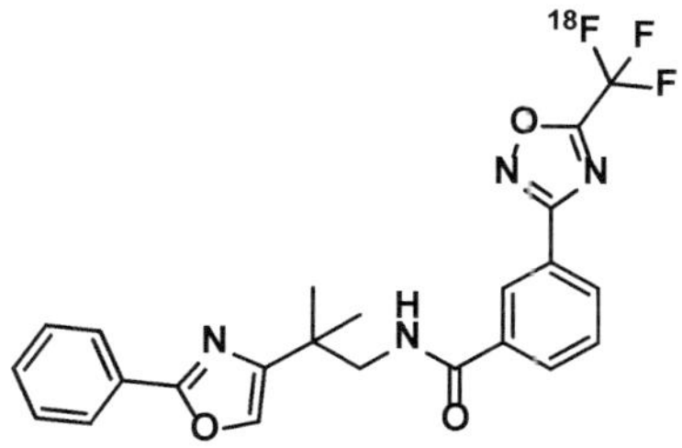

13

Structure 13

3. Oxadiazole-Based Probes for SPECT Imaging

Single-photon emission computed tomography (SPECT) is another non-invasive imaging technique used for mapping 3-dimensional images of organs, tissues, bones, etc. *In vivo* visualization of amyloid-β (Aβ) deposition in the brain facilitates early diagnosis of Alzheimer's disease (Chen et al. 2014). In this instance, SPECT is useful for visualizing amyloid pathology (Okumura et al. 2018). Additionally, SPECT imaging offers cheaper operating costs in comparison to PET imaging, which makes them suitable for primary screening of cerebral β-amyloidosis (Ono et al. 2013). Some of the iodinated (including $^{123/125}$I labeled) synthetic probes based on oxadiazole derivatives are discussed as follows.

In 2009, Watanabe et al. reported a series of iodine-attached 1,3,4-oxadiazole-based probes 14–19 (Watanabe et al. 2009). These probes exhibited good affinity toward Aβ42 aggregates (K_i ~ 20–349 nM). Excellent *in vitro* results were observed in mouse model with fluorescent staining of β-amyloid plaques.

In 2014, Watanabe et al. tailored oxadiazole probes 20–21 in three synthetic steps from commercially available 4-bromobenzoic hydrazide and 3-bromobenzoic hydrazide respectively (Watanabe et al. 2014). The developed iodinated probes 20–21 exhibited significant affinity (25 and 14 nM) toward Aβ (1–42) aggregates in *in vitro* experiments. Injection in the mouse revealed good penetration ability and fast washout. These probes, when labeled with ^{123}I, exhibited plaque localization owing to the specific binding of Aβ plaques.

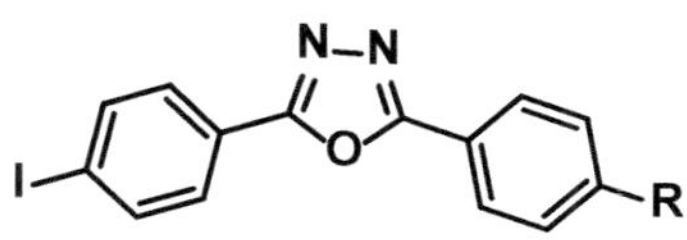

14: R = $N(CH_3)_2$
15: R = OCH_3
16: R = OH
17: R = OCH_2CH_2OH
18: R = $(OCH_2CH_2)_2OH$
19: R = $(OCH_2CH_2)_3OH$

Structures 14–19

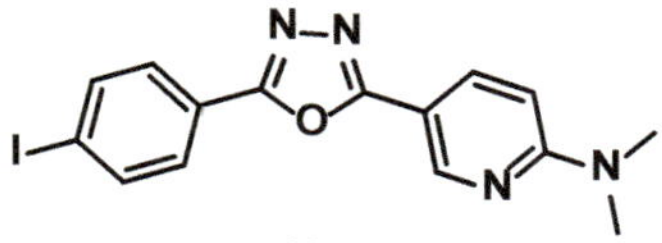

20

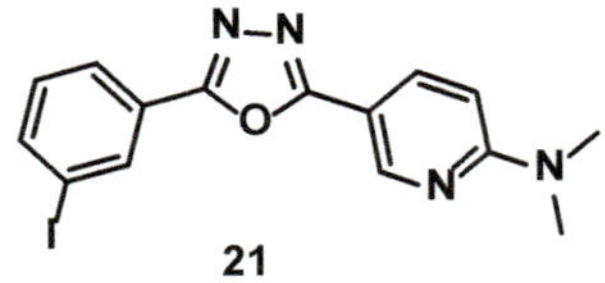

21

Structures 20 and 21

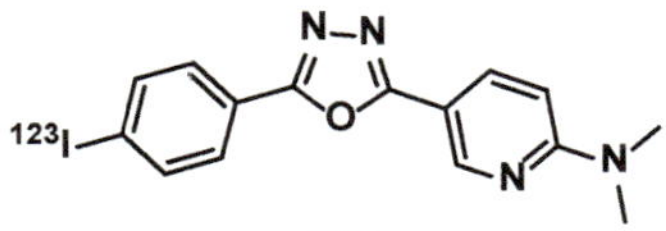

22a

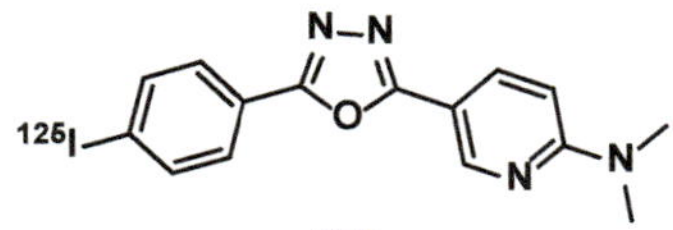

22b

Structures 22a and 22b

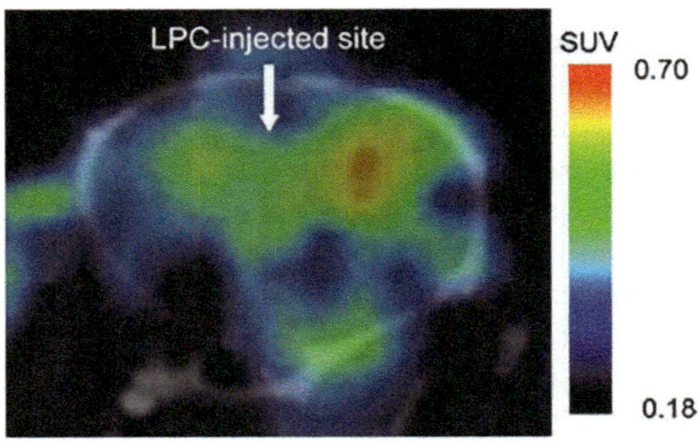

Figure 2. SPECT-CT imaging of mouse brain following LPC-induced demyelination over 15–60 min after treatment of probe 22a. Reproduced here with the permission of the American Chemical Society (Watanabe et al. 2022).

Myelin covers neuronal axons as a protective shield that facilitates neuronal signal conduction. Once the myelin is defaced because of neurodegenerative diseases, neuronal axons become more prone to damage. In 2022, Watanabe et al. also published an article (Watanabe et al. 2022). Probe 22b successfully detected myelin in the mouse brain after 1 h of injection. Further, demyelination was noticed in the presence of probe 22b in the brain and spinal cord of the mouse. Interestingly, demyelination (Figure 2) was noticed only in the brain on injection of probe 22a during SPECT imaging.

4. Oxadiazole-Based Probes as Enzyme Markers

Overexpression of cyclooxygenase-2 (COX-2) enzyme results in cancer-related health conditions. Also, the production of prostaglandins directly affects tissue invasion, metastasis carcinogenesis, etc. (De Vries 2006). Therefore, detection of COX-2 expression is important for early diagnosis and treatment. Kaur et al. developed a range of fluorescent probes 23–27 utilizing a click chemistry approach (Kaur et al. 2021). The azide derivative 23a is an excellent cell permeable intermediate, forming a COX-2 specific fluorescent probe through "in cell" click chemistry in the presence

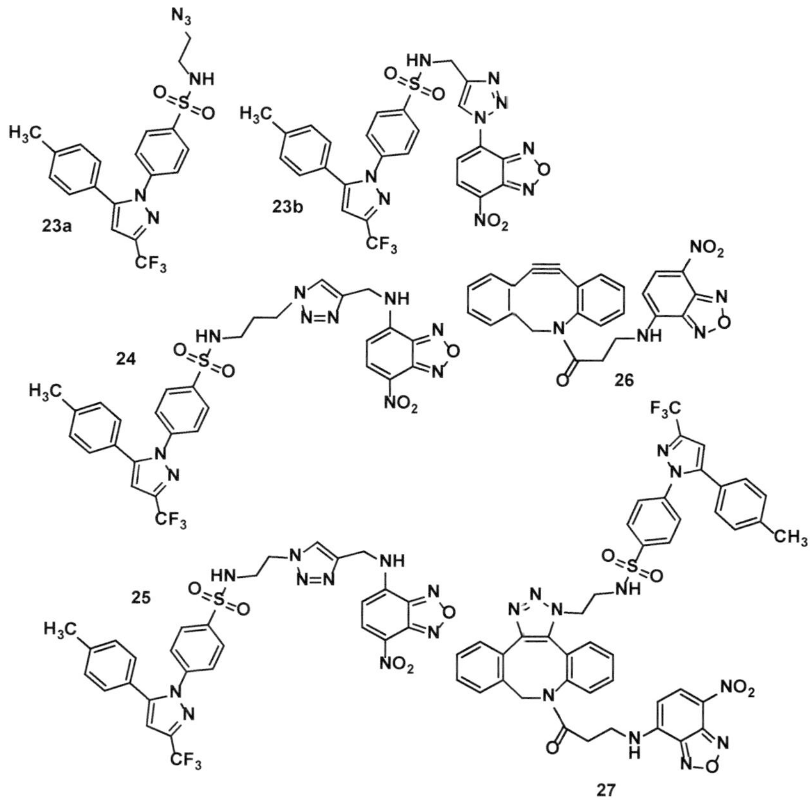

Structures 23–27

of compound 26. The probe detected COX-2 efficiently as a turn-on fluorescence signal in HCA-7 cells.

5. Oxadiazole-Based Probes in Protein Imaging

Fluorogenic probes are considered efficient markers owing to their spatial and temporal sensitivity. These are extensively used for monitoring protein translocation and imaging within living cells. Baranczak et al. studied the imaging response of a fluorosulfate-attached oxadiazole probe 28 (Baranczak et al. 2015). Efficient binding (2 nM) of the transthyretin (TTR) was observed when probe 28 was added to HEK293T cells. The probe could successfully penetrate live HEK293T cells and visualize exogenous transthyretin in the endoplasmic reticulum as well as the mitochondria compartment. Ono et al. developed a series of β-amyloid probes 29–33 that consist of a centrally located oxadiazole motif (Ono et al. 2008). ^{125}I labeled synthetic probes (29–32) exhibited remarkable uptake during imaging (2.2 to 3.3% ID/g). However, due to non-specific binding, these four ^{125}I labeled probes washed out slowly from the brain during *in vivo* studies. Further modification of the

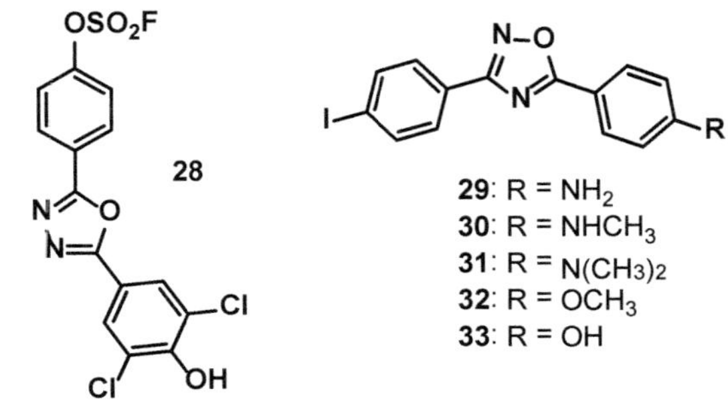

Structures 28–33

probe structure is required for fast washout. *In vitro* studies confirmed significant binding affinity aggregates ($K_i \sim 4$–47 nM) toward Aβ aggregates.

6. Nitric Oxide Imaging

In 2019, Han et al. developed a nitric oxide responsive fluorescent probe 34a utilizing a thiosemicarbazide derivative (Scheme 1) and a coumarin unit (Han et al. 2018). The synthetic probe displayed a fast turn-on fluorescence response in the presence of nitric oxide both in buffer solution and in live MCF-7 cells, and it finally produced the fluorescent probe 34b. The limit of detection during nitric oxide interaction was calculated to be 47.6 nM. The authors also demonstrated endogenous nitric oxide imaging in the zebrafish model using the probe 34a.

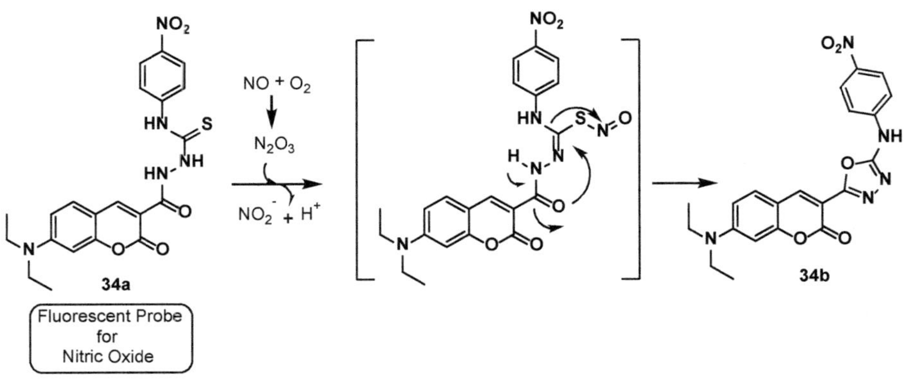

Scheme 1. Schematic representation of binding mechanism in probe 34.

7. Oxadiazole-Based Probes for Metal Ion Imaging

Oxadiazole is employed as an excellent metal ion chelator owing to the presence of suitable heteroatoms such as nitrogen and oxygen in its parent structure (Sahoo et al. 2017). Until recently, oxadiazole has been utilized in various metal ions imaging, including alkaline earth metals (Ca^{2+}), transition metal ions such as Fe^{2+}, and heavy metal ions such as Zn^{2+} and Cd^{2+}.

7.1 Al^{3+} Imaging

Manjunath et al. designed a receptor 35 with two rhodamine units attached through an oxadiazole moiety (Manjunath et al. 2015). The probe was isolated as a pale yellow solid in 80% synthetic yield. A distinct pink color, as well as pink fluorescence, appeared with the addition of Al^{3+} into a probe solution in aqueous acetonitrile. Excellent emission response was noticed at 568 nm upon adding Al^{3+} ions to the probe solution. The synthetic probe penetrated MDA-MB-231 breast cancer cells with no detectable fluorescence. However, injection of Al^{3+} resulted in red fluorescence during cell imaging experiments, indicating selective detection of Al^{3+} in live cells.

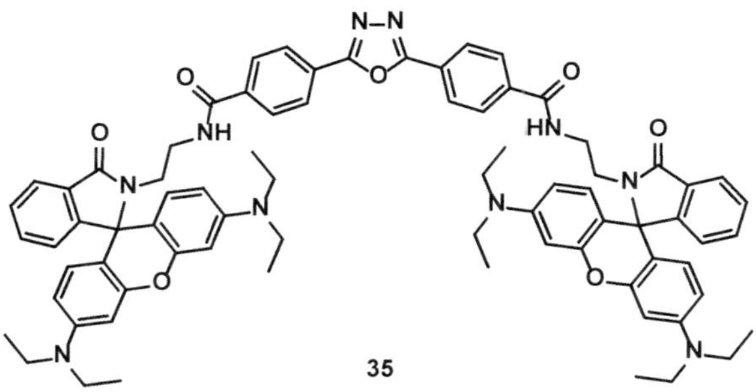

35

Structure 35

7.2 Ca^{2+} Imaging

Liu et al. developed a ratiometric fluorescent probe 36 utilizing BAPTA as a chelator for Ca^{2+} coordination (Scheme 2), an electron-deficient 1,3,4-oxadiazole unit, and an electron-rich ethoxy benzene (Liu et al. 2013). In the presence of increasing Ca^{2+} ion concentrations, the original UV-visible bands at 305 and 380 nm were reduced, whereas the band at 380 nm shifted to 350 nm with a distinct color change from orange to greenish yellow. The probe exhibited excellent selectivity toward Ca^{2+} ions with a very high Stokes shift of 202 nm owing to the presence of a conjugated system with a large π-electron—coordination of Ca^{2+}-induced emission spectral shifting from 582 nm to 490 nm with evolution of bright green fluorescence. Interestingly, the probe was utilized to detect intracellular Ca^{2+} fluctuation in live HUVEC (human umbilical vein endothelial) cells.

7.3 $Fe^{2+/3+}$ Imaging

Gong et al. developed a fluorescent probe 38, which consists of pyridine carboxylic acid moieties and a 1,3,4-oxadiazole unit (Gong et al. 2019). The fluorescence response (intensity at 354 nm) of the probe was quenched significantly upon the addition of Fe^{2+} and Fe^{3+} in aqueous solution. A 1:1 Binding coordination was observed between the oxadiazole probe and the iron species. A low limit of detection was envisaged, particularly 6.95 μM for Fe^{3+} and 7.78 μM for Fe^{2+}, respectively,

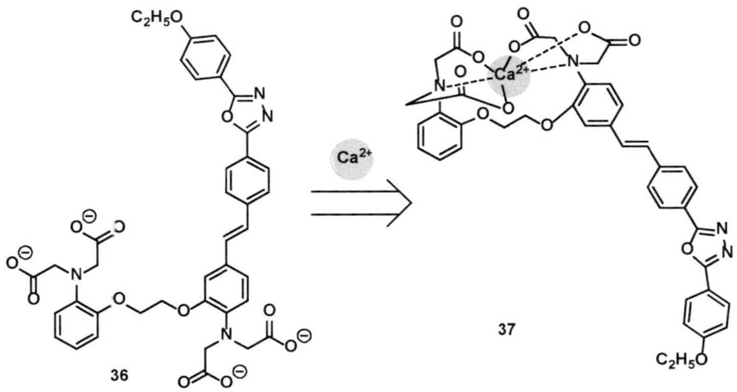

Scheme 2. Mechanism of binding of calcium ion to probe 36 to produce a complex 37.

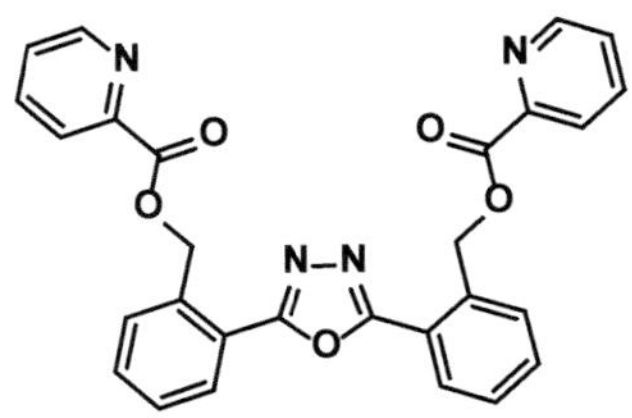

38
Structure 38

during metal coordination. The fluorescent probe exhibited good cell permeability in HepG2 cells.

7.4 Zn^{2+} Imaging

The existence of potential coordination sites such as nitrogen and oxygen atoms makes these oxadiazole derivatives notable scaffolds for metal ion binding. Zhou et al. developed a Zn^{2+} imaging tool where the synthetic probe 39 is made up of 2-hydroxy-phenyl-1,3,4-oxadiazole as a chelating site (Zhou et al. 2012). The probe successfully detected Zn^{2+} ions in aqueous CH_3CN by following 1:1 binding coordination. The addition of Zn^{2+} resulted in a dramatic enhancement of fluorescence intensity at 439 nm (~ 65 fold) with an association constant of 1.6×10^5 M^{-1}. Significant fluorescence response was noticed from fluorescence microscopy studies of probe 39 in the presence of Zn^{2+} in liver cancer (Hep G2) cells owing to excellent cell penetration ability and electrophilic attack of zinc ion to the oxygen atom of phenol unit.

Tang et al. prepared a fluorescent probe 40 in 72% yield utilizing 2,5-diphenyl-1,3,4-oxadiazole scaffold and dipicolylamine unit (Tang et al. 2014). It detected Zn^{2+} selectively following an ESIPT (excited state intramolecular proton transfer) mechanism in pure water at physiological pH. The Zn^{2+} coordination followed 1:1 binding stoichiometry with a red-shifted emission enhancement at

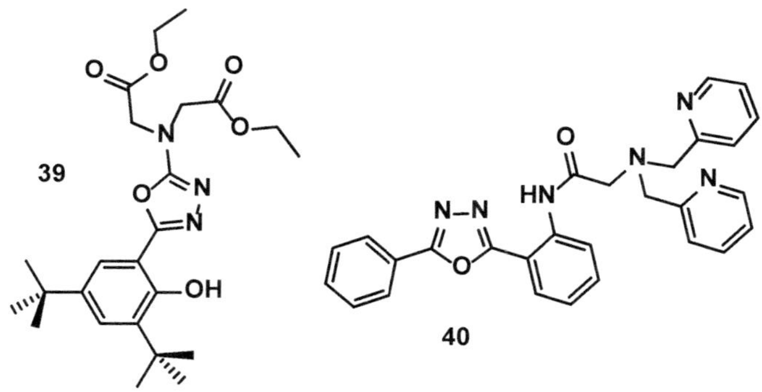

Structures 39 and 40

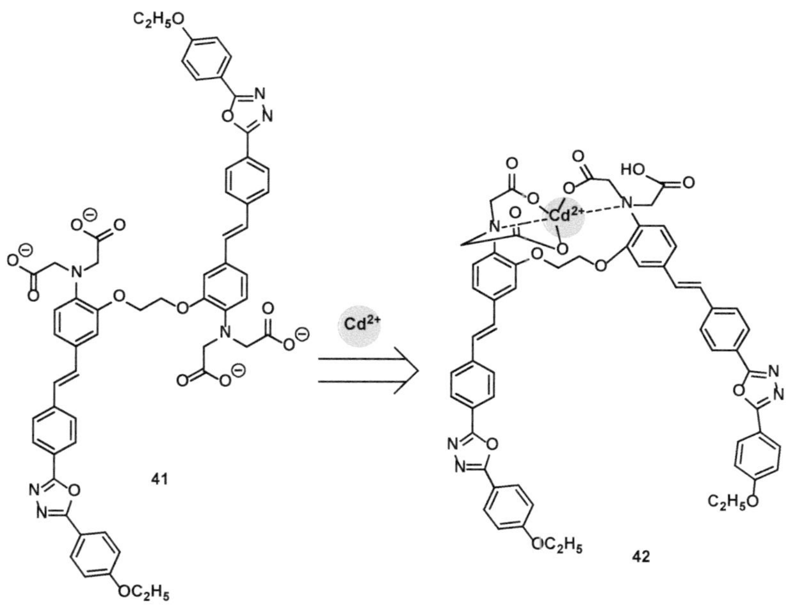

Scheme 3. Mechanism of binding of cadmium ion to probe 41 to produce a complex (42).

440 nm. The evolution of green fluorescence from live HeLa cells in the presence of Zn^{2+} ions confirmed cell permeability and Zn^{2+} sensing characteristics of the fluorescent probe.

Liu et al. reported a BAPTA [1,2-bis(*o*-aminophenoxy)ethane-N,N,N′,N′-tetraacetic acid] based fluorescent probe 41 with two oxadiazole units (Liu et al. 2014). The synthesized probe selectively detected Cd^{2+} in an aqueous medium with 20 nM as the detection limit (Scheme 3). The probe might be useful for sensitive detection of Cd^{2+}, which also exhibited a large Stokes shift of 189 nm. The esterified probe exhibited good cell penetration and detected Cd^{2+} with bright emission in MCF-7 cells.

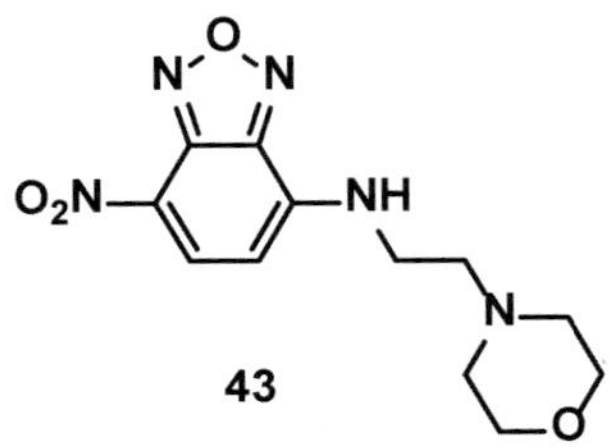

43

Structure 43

Zhang et al. developed a small molecule probe 43 where 4-nitro-benzo[1,2,5] oxadiazole is covalently attached to a morpholine moiety through an aminoethylene bridge (Zhang et al. 2016). The synthesized fluorescent probe 43 detected Hg^{2+} in live HeLa cells and exhibited better biocompatibility and cell-permeable characteristics. Single crystal x-ray studies revealed the 2:2 binding geometry of Hg^{2+} with the probe, where Hg^{2+} is held together by nitrogen atoms of morpholine and the amino group.

8. Oxadiazole-Based Probes for pH Imaging

Huang et al. developed a carboxylic acid-based pH activatable probe 44 in 45% yield utilizing folate residue with 4-nitro-benzo[1,2,5]oxadiazole fluorophore attached piperazine precursor (Huang et al. 2012). The developed probe exhibited remarkable fluorescence enhancement (~ 30 fold) under an acidic environment (pK_a = 5.70). Interestingly, it was able to respond to pH changes in cancerous live cells and might be useful for mapping intracellular pH distributions.

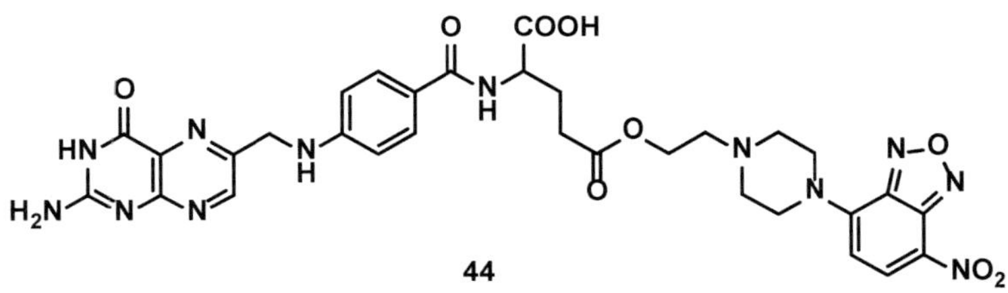

44

Structure 44

9. Oxadiazole-Based Probes in Cell Imaging and Nrf2 Activation

Oxadiazole-based probes have been utilized in various ways in disease prevention and cell biology. Due to its significant therapeutical efficacy, oxadiazole motifs are used in brain tissue targeting (Xu et al. 2019), Nrf2 activation (Xu et al. 2018), and neuro-protective agents (Lin et al. 2020). The nuclear factor erythroid 2-related factor 2 (Nrf2) is widely accepted as the master regulator of an antioxidant defense system. Nrf2 translocation in cell organelles can be monitored using oxadiazole derivatives (Xu et al. 2016). Some recent applications of oxadiazole probes in cell imaging studies are highlighted below.

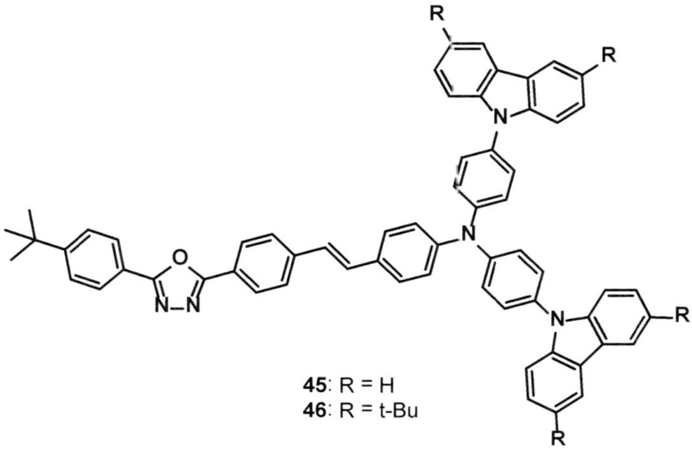

Structures 45 and 46

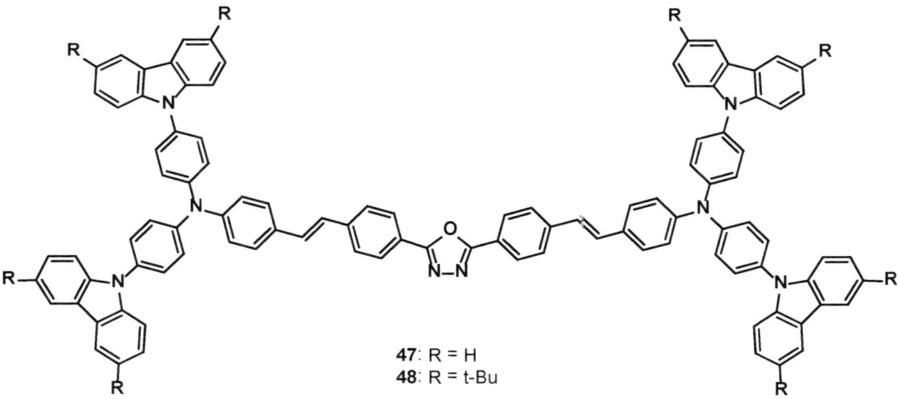

Structures 47 and 48

Jin et al. tailored carbazole-decorated conjugated fluorescent probes 45–48 attached with triphenylamine and 1,3,4-oxadiazole unit (Jin and Qian 2015). Spectroscopic data from a mixed solvent of water-THF suggested intramolecular charge transfer as well as aggregation-induced emission behavior. Fabrication of dye-loaded silica nanoparticles improved biocompatibility, and at the same time, dye-loaded Bovine Serum Albumin (BSA) nanoparticles exhibited excellent uptake by HeLa cells in cell imaging studies.

Li et al. synthesized a triphenylamine-centered conjugated probe 49 attached with a BODIPY fluorophore using a palladium-catalyzed Heck coupling reaction in 54.8% yield (Li and Qian 2017). The probe displayed an excellent red fluorescence in the solid state (λ_{em} = 614 nm). The THF solution exhibited two emission bands (647 and 513 nm) with fluorescence quantum yields of 0.11 and 0.01, respectively. The fluorescence intensity of the probe enhanced in the presence of an increasing concentration of Bovine Serum Albumin (BSA), indicating a BSA selective

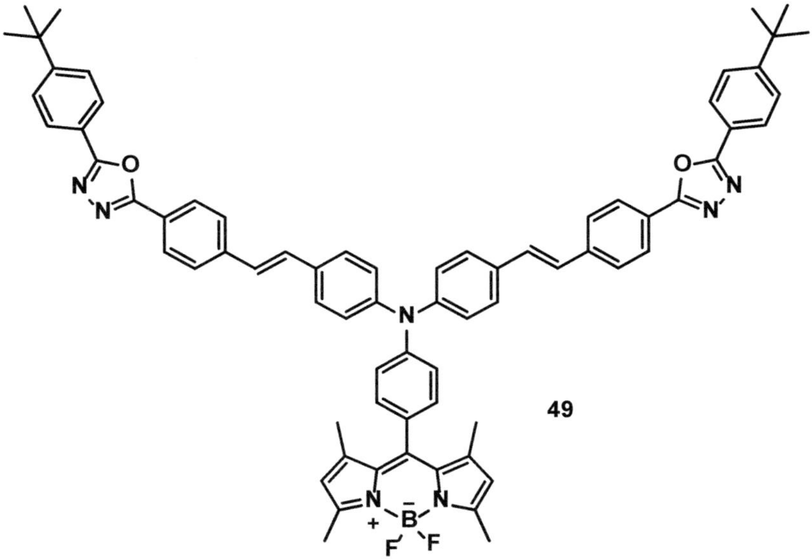

49

Structures 49

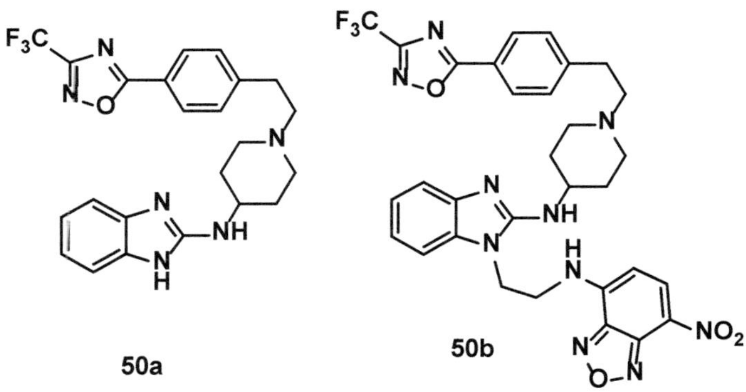

50a

50b

Structures 50a and 50b

fluorescent probe. Confocal fluorescence microscopy studies revealed good cell uptake by MDA-MB-231 cells using the synthesized probe 49.

Mambwe et al. optimized an ester containing astemizole (a second-generation antihistamine) compound using an iterative medicinal chemistry approach and obtained a highly bioactive candidate 50a (Mambwe et al. 2022). Probe 50a was tested for its efficacy studies, which displayed strong *in vitro* activity with an IC50 (PfNF54) value 0.012 μM w.r.t drug-sensitive strain of P. falciparum. It also displayed excellent *in vivo* potency in the mouse model. In order to obtain live cell imaging results, the authors have installed an extrinsic fluorophore such as nitrobenzoxadiazole (NBD) to the parent structure to yield 50b as a fluorescent probe. A significant accumulation of probe 50b was observed around hemozoin crystals in super-resolution structured illumination microscopy images.

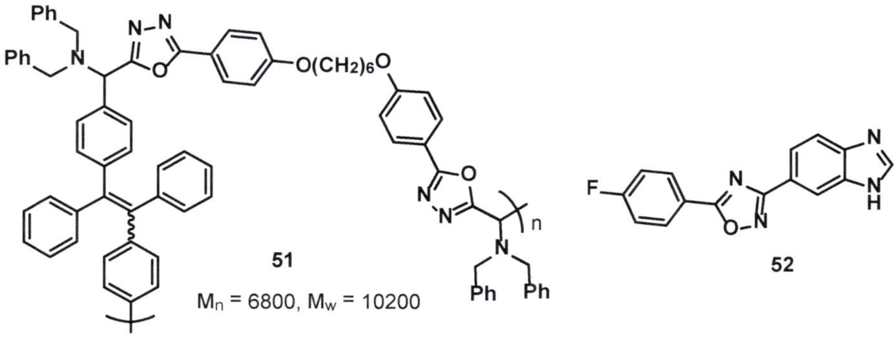

Structures 51 and 52

Xie et al. prepared polyoxadiazole derivative 51 in 61% yield through a multicomponent polymerization process (catalyst-free), which ensured better film-forming capacity, good solubility, and high thermal stability (Xie et al. 2023). The polymerized probe exhibited excellent solid-state fluorescence response due to the presence of aggregation-induced emission part in the host system. The probe also exhibited bright blue fluorescence in 4T1 cells, which confirmed good cell staining potential and was found to be selective toward lysosomes.

In order to obtain the best possible Nrf2 activators, Xu et al. screened 7,500 compounds and obtained a very potent probe 52, which displayed the best cell permeability as well as good solubility (Xu et al. 2015). Pharmacological investigations such as *in vivo* studies revealed that probe 52 effectively suppressed inflammation in LPS-challenged mice. Nucleus translocation of Nrf2 was also visualized using an immunofluorescence assay (Figure 3). Strong fluorescence signal

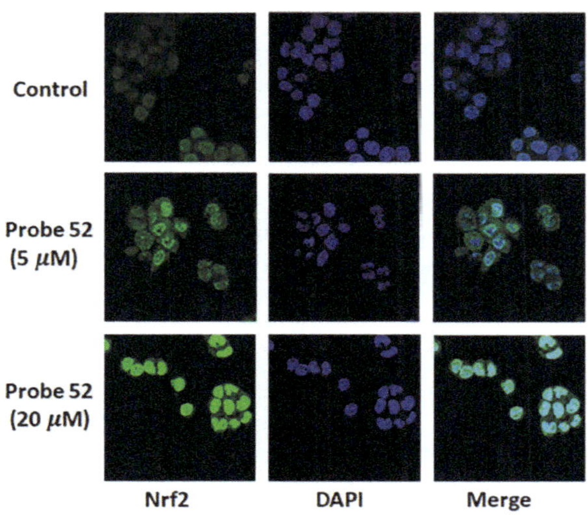

Figure 3. Confocal microscopy images of Nrf2 and nucleus in HCT116 cells. Green staining (labeled with DyLight 488) shows Nrf2, blue staining (labeled with DAPI) signifies nuclei and merge represents both. Reproduced here with the permission of the American Chemical Society (Xu et al. 2015).

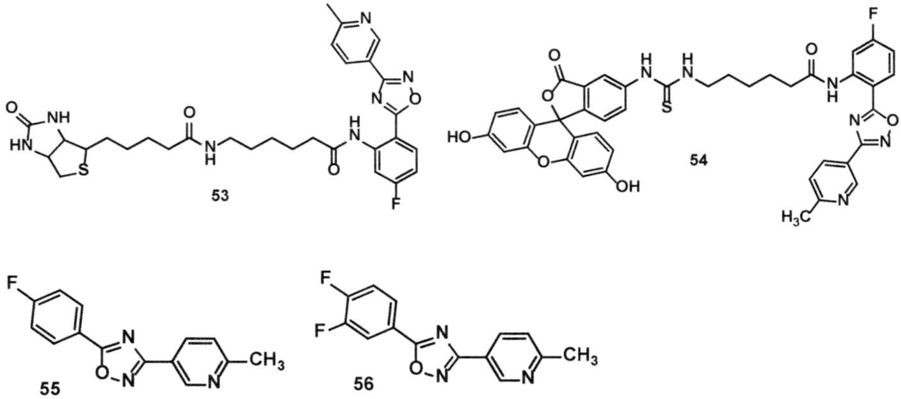

DDO-7263

Structures 53–56

in the nucleus from confocal images shows that Nrf2 translocated into the nucleus in a concentration-dependent fashion.

The Nrf2 protein plays a significant role in various enzyme expressions as well as antioxidant protein activation. Dai et al. developed several oxadiazole-based probes 53–55 (Dai et al. 2022). Compound 53 is a biotin-based probe, and compound 54 is a fluorescent probe designed for Nrf2 activation purposes. These synthesized probes exhibited excellent antioxidant as well as anti-inflammatory responses similar to the commercially available Nrf2-ARE activator DDO-7263 (56). DDO-7263 mimic is based on a 1,2,4-oxadiazole positioned centrally and attached to difluorobenzene moiety in one end and methyl pyridine in the other end. It has good anti-inflammatory and Nrf2-activating properties. The binding affinity assay studies revealed that the synthesized probes displayed antioxidant behavior owing to the binding to a component of 26S proteasome called RPN6. The studies confirmed that the 1,2,4-oxadiazole derivatives promoted Nrf2 activation.

10. Conclusions

In this chapter, we have elucidated diverse sets of oxadiazole-based synthetic probes applicable in cell imaging applications. Although oxadiazole-based motifs are principal precursors in most drug candidates used in medicinal chemistry, few synthetic probes have been studied for live-cell imaging purposes. Therefore, a rational design of small molecule probes with structural diversity must be developed, including various subcellular targeting groups and tunable photophysical behavior. In the meantime, oxadiazole has been functionalized with different fluorophores such as quinoline, coumarin, rhodamine, di-(2-picolyl)amine, carbazole, BODIPY, etc., and with different chelating groups such as BAPTA. Besides that, oxadiazole has also been functionalized with targeting groups such as morpholine for selective targeting inside the lysosome compartment. The cell penetration efficiency of the synthesized probes has been discussed in various cell lines.

References

Ahamed, M., D. Van Veghel, C. Ullmer, K. Van Laere, A. Verbruggen and G. M. Bormans. 2016. Synthesis, biodistribution and *in vitro* evaluation of brain permeable high affinity type 2 cannabinoid receptor agonists [11C] MA2 and [18F] MA3. Front. Neurosci. 10: 431.

Attili, B., S. Celen, M. Ahamed, M. Koole, C. V. D. Haute, W. Vanduffel et al. 2019. Preclinical evaluation of [18F] MA3: a CB2 receptor agonist radiotracer for PET. Brit. J. Pharmacol. 176(10): 1481–1491.

Baranczak, A., Y. Liu, S. Connelly, W.-G. H. Du, E. R. Greiner, J. C. Genereux et al. 2015. A fluorogenic aryl fluorosulfate for intraorganellar transthyretin imaging in living cells and in Caenorhabditis elegans. J. Am. Chem. Soc. 137(23): 7404–7414.

Chen, C.-J., K. Bando, H. Ashino, K. Taguchi, H. Shiraishi, K. Shima et al. 2014. Biological evaluation of the radioiodinated imidazo [1, 2-a] pyridine derivative DRK092 for amyloid-β imaging in mouse model of Alzheimer's disease. Neurosci. Lett. 581: 103–108.

Dai, Z., L.-y. An, X.-y. Chen, F. Yang, N. Zhao, C.-c. Li et al. 2022. Target fishing reveals a novel mechanism of 1, 2, 4-oxadiazole derivatives targeting rpn6, a subunit of 26S proteasome. J. Med. Chem. 65(6): 5029–5043.

De Vries, E. 2006. Imaging of cyclooxygenase-2 (COX-2) expression: potential use in diagnosis and drug evaluation. Curr. Pharm. Des. 12(30): 3847–3856.

Gazzi, T., B. Brennecke, K. Atz, C. Korn, D. Sykes, G. Forn-Cuni et al. 2022. Detection of cannabinoid receptor type 2 in native cells and zebrafish with a highly potent, cell-permeable fluorescent probe. Chem. Sci. 13(19): 5539–5545.

Gong, X., H. Zhang, N. Jiang, L. Wang and G. Wang. 2019. Oxadiazole-based 'on-off' fluorescence chemosensor for rapid recognition and detection of Fe^{2+} and Fe^{3+} in aqueous solution and in living cells. Microchem. J. 145: 435–443.

Han, Q., J. Liu, Q. Meng, Y.-L. Wang, H. Feng, Z. Zhang et al. 2018. Turn-on fluorescence probe for nitric oxide detection and bioimaging in live cells and zebrafish. ACS Sens. 4(2): 309–316.

Huang, R., S. Yan, X. Zheng, F. Luo, M. Deng, B. Fu et al. 2012. Development of a pH-activatable fluorescent probe and its application for visualizing cellular pH change. Analyst 137(19): 4418–4420.

Hughes, G. and M. R. Bryce. 2005. Electron-transporting materials for organic electroluminescent and electrophosphorescent devices. J. Mater. Chem. 15(1): 94–107.

Jin, Y. and Y. Qian. 2015. Photophysical properties, aggregation-induced fluorescence in nanoaggregates and cell imaging of 2, 5-bisaryl 1, 3, 4-oxadiazoles. New J. Chem. 39(4): 2872–2880.

Kaur, J., A. Bhardwaj and F. Wuest. 2021. In Cellulo Generation of Fluorescent Probes for Live-Cell Imaging of Cylooxygenase-2. Chem. Eur. J. 27(10): 3326–3337.

Kumata, K., J. Yui, L. Xie, Y. Zhang, N. Nengaki, M. Fujinaga et al. 2015. Radiosynthesis and preliminary PET evaluation of glycogen synthase kinase 3β (GSK-3β) inhibitors containing [11C] methylsulfanyl,[11C] methylsulfinyl or [11C] methylsulfonyl groups. Bioorg. Med. Chem. Lett. 25(16): 3230–3233.

Li, Q. and Y. Qian. 2017. A red-emissive oxadiazol-triphenylamine BODIPY dye: synthesis, aggregation-induced fluorescence enhancement and cell imaging. J. Photochem. Photobiol., A: Chem. 336: 183–190.

Lin, H., Y. Qiao, H. Yang, Q. Li, Y. Chen, W. Qu et al. 2020. Design and evaluation of Nrf2 activators with 1, 3, 4-oxa/thiadiazole core as neuro-protective agents against oxidative stress in PC-12 cells. Bioorg. Med. Chem. Lett. 30(2): 126853.

Liu, Q., W. Bian, H. Shi, L. Fan, S. Shuang, C. Dong et al. 2013. A novel ratiometric emission probe for Ca^{2+} in living cells. Org. Biomol. Chem. 11(3): 503–508.

Liu, Q., L. Feng, C. Yuan, L. Zhang, S. Shuang, C. Dong et al. 2014. A highly selective fluorescent probe for cadmium ions in aqueous solution and living cells. Chem. Commun. 50(19): 2498–2501.

Luo, Z., J. Han, H. Liu, A. J. Rosenberg, D. L. Chen, R. J. Gropler et al. 2018. Syntheses and *in vitro* biological evaluation of S1PR1 ligands and PET studies of four F-18 labeled radiotracers in the brain of nonhuman primates. Org. Biomol. Chem. 16(47): 9171–9184.

Luo, Z., A. J. Rosenberg, H. Liu, J. Han and Z. Tu. 2018. Syntheses and *in vitro* evaluation of new S1PR1 compounds and initial evaluation of a lead F-18 radiotracer in rodents. Eur. J. Med. Chem. 150: 796–808.

Maan, A., R. S. Mathpati, V. D. Ghule and S. Dharavath. 2021. Effect of multiple oxadiazole rings with nitro and nitramino functionalities on energetic properties: computational analysis of the structure–property relationship. New J. Chem. 45(16): 7368–7376.

Mambwe, D., C. M. Korkor, A. Mabhula, Z. Ngqumba, C. Cloete, M. Kumar et al. 2022. Novel 3-Trifluoromethyl-1, 2, 4-oxadiazole analogues of astemizole with multi-stage antiplasmodium activity and *in vivo* efficacy in a Plasmodium berghei mouse malaria infection model. J. Med. Chem.

Manjunath, R., E. Hrishikesan and P. Kannan. 2015. A selective colorimetric and fluorescent sensor for Al^{3+} ion and its application to cellular imaging. Spectrochim. Acta A Mol. Biomol. Spectrosc. 140: 509–515.

Mishra, R. 2010. Glycogen synthase kinase 3 beta: can it be a target for oral cancer. Mol. Cancer 9: 1–15.

Okumura, Y., Y. Maya, T. Onishi, Y. Shoyama, A. Izawa, D. Nakamura et al. 2018. Design, synthesis, and preliminary evaluation of SPECT probes for imaging β-Amyloid in Alzheimer's disease affected brain. ACS Chem. Neurosci. 9(6): 1503–1514.

Ono, M., Y. Cheng, H. Kimura, H. Watanabe, K. Matsumura, M. Yoshimura et al. 2013. Development of novel 123I-labeled pyridyl benzofuran derivatives for SPECT imaging of β-amyloid plaques in Alzheimer's disease. PloS One 8(9): e74104.

Ono, M., M. Haratake, H. Saji and M. Nakayama. 2008. Development of novel β-amyloid probes based on 3, 5-diphenyl-1, 2, 4-oxadiazole. Bioorg. Med. Chem. 16(14): 6867–6872.

Ropero, S. and M. Esteller. 2007. The role of histone deacetylases (HDACs) in human cancer. Mol. Oncol. 1(1): 19–25.

Rosenberg, A. J., H. Liu, H. Jin, X. Yue, S. Riley, S. J. Brown et al. 2016. Design, synthesis, and *in vitro* and *in vivo* evaluation of an 18F-labeled sphingosine 1-phosphate receptor 1 (S1P1) PET tracer. J. Med. Chem. 59(13): 6201–6220.

Sahoo, P. R., K. Prakash and S. Kumar. 2017. Synthesis of an oxadiazole through an indole mediated single step procedure for selective optical recognition of Cu^{2+} ions. Sens. Actuators B Chem. 242: 299–304.

Siwach, A. and P. K. Verma. 2020. Therapeutic potential of oxadiazole or furadiazole containing compounds. BMC Chem. 14: 1–40.

Tang, L., X. Dai, K. Zhong, D. Wu and X. Wen. 2014. A novel 2, 5-diphenyl-1, 3, 4-oxadiazole derived fluorescent sensor for highly selective and ratiometric recognition of Zn^{2+} in water through switching on ESIPT. Sens. Actuators B: Chem. 203: 557–564.

Turkman, N., D. Liu and I. Pirola. 2021. Novel late-stage radiosynthesis of 5-[18F]-trifluoromethyl-1, 2, 4-oxadiazole (TFMO) containing molecules for PET imaging. Sci. Rep. 11(1): 1–10.

Watanabe, H., R. Maekawa, S. Iikuni, M. Kakae, N. Matsuo, H. Shirakawa et al. 2022. Characterization of radioiodinated diaryl oxadiazole derivatives as SPECT probes for detection of myelin in multiple sclerosis. ACS Chem. Neurosci. 13(3): 363–369.

Watanabe, H., M. Ono, R. Ikeoka, M. Haratake, H. Saji and M. Nakayama. 2009. Synthesis and biological evaluation of radioiodinated 2, 5-diphenyl-1, 3, 4-oxadiazoles for detecting β-amyloid plaques in the brain. Bioorg. Med. Chem. 17(17): 6402–6406.

Watanabe, H., M. Ono, H. Kimura, K. Matsumura, M. Yoshimura, S. Iikuni et al. 2014. Novel radioiodinated 1, 3, 4-oxadiazole derivatives with improved *in vivo* properties for SPECT imaging of β-amyloid plaques. MedChemComm. 5(1): 82–85.

Xie, J., N. Niu, X. Fu, X. Su, D. Wang, A. Qin et al. 2023. Catalyst-free synthesis of diverse fluorescent polyoxadiazoles for the facile formation and morphology visualization of microporous films and cell imaging. Chem. Sci.

Xu, L.-L., Y.-F. Wu, L. Wang, C.-C. Li, L. Li, B. Di et al. 2018. Structure-activity and structure-property relationships of novel Nrf2 activators with a 1, 2, 4-oxadiazole core and their therapeutic effects on acetaminophen (APAP)-induced acute liver injury. Eur. J. Med. Chem. 157: 1376–1394.

Xu, L.-L., Y.-F. Wu, F. Yan, C.-C. Li, Z. Dai, Q.-D. You et al. 2019. 5-(3, 4-Difluorophenyl)-3-(6-methylpyridin-3-yl)-1, 2, 4-oxadiazole (DDO-7263), a novel Nrf2 activator targeting brain tissue, protects against MPTP-induced subacute Parkinson's disease in mice by inhibiting the NLRP3 inflammasome and protects PC12 cells against oxidative stress. Free Radic. Biol. Med. 134: 288–303.

Xu, L.-L., X. Zhang, Z.-Y. Jiang and Q.-D. You. 2016. Molecular similarity guided optimization of novel Nrf2 activators with 1, 2, 4-oxadiazole core. Bioorg. Med. Chem. 24(16): 3540–3547.

Xu, L.-L., J.-F. Zhu, X.-L. Xu, J. Zhu, L. Li, M.-Y. Xi et al. 2015. Discovery and modification of *in vivo* active Nrf2 activators with 1, 2, 4-oxadiazole core: hits identification and structure–activity relationship study. J. Med. Chem. 58(14): 5419–5436.

Zhang, Y., H. Chen, D. Chen, D. Wu, Z. Chen, J. Zhang et al. 2016. A colorimetric and ratiometric fluorescent probe for mercury (II) in lysosome. Sens. Actuators B: Chem. 224: 907–914.

Zhou, J.-A., X.-L. Tang, J. Cheng, Z.-H. Ju, L.-Z. Yang, W.-S. Liu et al. 2012. An 1, 3, 4-oxadiazole-based OFF–ON fluorescent chemosensor for Zn^{2+} in aqueous solution and imaging application in living cells. Dalton Trans. 41(35): 10626–10632.

CHAPTER 7

Oxadiazole as Inhibitors

1. Introduction

The development of antibiotics has significantly increased life expectancy. Antibiotics are the main hope for saving millions of lives from infections, which is believed to be of utmost importance. However, the danger from several drug-resistant bacterial infections is a serious threat to humankind (Bassetti et al. 2017). The chemical and nature-borne molecules that contain heterocyclic compounds are widely recognized as ligands, intermediates, inhibitors, agrochemicals, and veterinary treatments. They exhibit strong biological action against bacterial infections. In this context, oxadiazole and its derivatives, such as 1,2,4-oxadiazole, 1,3,4-oxadiazole, and 1,2,5-oxadiazole, are widely recognized as the most potent inhibitors against drug-resistant bacteria (Naclerio et al. 2019, 2020) because they are approved by many pharmaceuticals which encourage chemists to use them in the fields of medicinal chemistry, polymer and material science, and pesticide chemistry due to their ability to mimic esters and amides as bioisosteres (Verma et al. 2021a).

The clinical success of β-lactam antibiotics over the past few decades creates numerous prospects for developing novel penicillin-binding proteins (PBPs) inhibitors (Frimodt-Meller et al. 1986; Ford et al. 1996). Since PBPs are essential for the biosynthesis of cell walls, they are still prime targets for antibiotics. Since biofilms are responsible for almost 65% of current human bacterial infections, they are also a significant target for medication. The biofilm will be the main cause of bacterial infections in the near future (Verma et al. 2021a). Methicillin-resistant Staphylococcus aureus (MRSA) infection, as noted by the Centers for Disease Control and Prevention (CDC) in its 2013 report, causes high morbidity and mortality (Verma et al. 2021a). In order to combat such infections, oxazolidinones are easily accessible oral drugs that are highly recommended over licensed drugs such as vancomycin, telavancin, daptomycin, ceftaroline, linezolid, and tedizolid (Kaka et al. 2006). Oxadiazole isosters are a superior structural alteration for oxazolidinones. The oxadiazole moiety is a significant structural motif in bioorganic and pharmaceutical chemistry, as well as a fundamental building block in organic chemistry (Salassa and Terenzi 2016). Due to its therapeutic potential in numerous medicinal applications and simplicity of preparation, it attracts a great deal of interest from a wide range of researchers. Due to their enhanced resistance to esterases' hydrolytic cleavage, oxadiazole is crucial in the bioisosterism of esters and amides (Verma et al. 2021b).

Importantly, oxadiazole and its derivatives are attracting the most attention as the preferable anticancer medications. Since current anticancer medications inhabit many enzymes and growth regulators that control cell development, their effects are carried out by suppressing cells (Matore et al. 2022). According to numerous studies, the ideal target for developing anticancer medications and preventing metastasis is the suppression of angiogenesis (Carmeliet et al. 2005). The histone proteins play crucial roles in regulating gene expression by altering numerous chemical events, like acetylation, phosphorylation, and methylation. Histone deacetylase (HDAC) inhibitors can regulate genes and ultimately exert control over the numerous elements essential for the growth and spread of cancer (Bi et al. 2006; Eckschlager et al. 2017).

This chapter provides a thorough overview of the structure-activity relationship (SAR) of oxadiazole scaffolds as inhibitors against MRSA strains and therapeutic potential on various HDAC enzymes and anticancer activity.

2. The Derivatized Compound From Oxadiazole: A Potential Inhibitor

2.1 Inhibitors as Anti-Methicillin-Resistant Staphylococcus aureus (MRSA) Agents

The fundamental structural difference between pathogenic bacteria and healthy cells provides significant insights for creating novel antibiotic substances. Bacterial infections are becoming harder to recover, necessitating the development of new, attractive strategies for infection control. In order to repair and regenerate soft tissue in wounds, a variety of antibacterial wound-healing hydrogels have been used, including hydrogels with inorganic antibacterial chemicals. The key drawbacks of the current inorganic hydrogels are their cytotoxicity, lack of sensitive non-degradability, and crucial developments. The oxadiazole ring, a potent inhibitor, is frequently utilized to treat diverse diseases; structural modification of this moiety to combat MRSA by using various substituents to increase metabolic stability is extensively researched (Manukumar et al. 2017).

Amphiphilic cell antimicrobials using oxadiazole-moiety-incorporated poly(-caprolactone)-poly (-ethylene glycol)-poly(-caprolactone) (PCEC-QAS) 13 have been studied to treat the MRSA-infected lesion. They are efficient against MRSA and *E. coli* both *in vitro* and *in vivo*, mainly due to their remarkable thermal stability up to 50°C. PCEC-QAS promotes wound healing without using outside medications and grows thick epidermis on impaired skin wounds like normal skin. This makes it an emerging skin-regenerating material that prevents bacterial infection. This has been confirmed by the Fluorescence-activated cell sorting (FACS) analysis, which shows that changes in the membrane potential directly impact the cytoplasmic components because QAS electrostatic interactions with bacteria cause the MRSA cell membrane damage. The absence of the oxadiazole ring in PCEC-QAS is thought to be the cause of its non-cytotoxicity rather than the presence of hydrocarbon chains, which could cause cytotoxicity due to their capacity to pass through cell membranes. The 3T3 cells' cytocompatibility with PCEC-QAS demonstrated good cell proliferation. The *in vivo* studies give a clear understanding of hydrogel degradation, which is required for any antibacterial agents to become non-toxic (Liu et al. 2020).

By developing the thiazol-oxadiazole core with various lipophilic, hydrophilic, lipophobic, and hydrophobic structural units, the antibacterial activity and pharmacokinetic properties are improved. Several oxadiazole derivatives demonstrated considerable activity against MRSA (2658 RCMB), with MIC values ranging from 1.95 to 3.90 mg/mL. Vancomycin was equally effective against compounds with guanidine and piperazine moieties with a MIC of 1.95 mg/mL. The SAR demonstrated that longer chain/sterically hindered groups on amines with aliphatic alkyl or alicyclic portion significantly decrease the activity against MRSA compared to shorter amines with less than four carbon atoms. Interestingly, MRSA was successfully combated by amines with polar attachments, such as guanidine, piperazine, hydroxyazetidine, and morpholine, respectively (Hannoun et al. 2020). Different heterocycle conjugates have recently been investigated as possible scaffolds with good selectivity and efficiency against various biological targets (Chen et al. 2017; Ravindar et al. 2018). Similarly, oxadiazole showed increased resistance to MRSA after conjugation with thiophene heterocycle. To create a SAR, the oxadiazole core has been combined with a polysubstituted thiophene ring. *S. aureus, P. auregenosa, B. pumilus* and *E. coli* were used as test organisms to determine the antibacterial potency. The results showed a strong inhibition with the 4-fluorophenyl group, which was highly encouraging. This might result from fluorine's medicinal effects and its metabolic stability. Studies on growth kinetics demonstrated the superior inhibitory capacity against *S. aureus* (Singla et al. 2020). On the oxadiazole core, various aryl groups added at the second and fifth carbons were effective against a variety of bacterial strains, including MRSA 16. The SAR demonstrated that the bromo group is suitable for preventing the growth of bacterial strains in the ortho position of the benzene ring in the second and fifth positions of the oxadiazole moiety. While other functional groups, such as nitro and methyl, do not exhibit significant action against any of the tested strains, the chlorine substitution on the aryl ring demonstrated modest activity. The compound's greater activity may be due to the bromine atom's bulkiness, which can adhere to microbe cells and cause apoptosis (Zabiulla et al. 2020). The oxadiazole derivates showing antibacterial activity are presented in Figure 1.

Many Gram-positive bacteria, including *S. aureus*, can thrive and reproduce due to teichoic acid and lipoteichoic acid (LTA) polymers that make up the peptidoglycan layer of Gram-positive bacteria anchored with teichoic acid. LTA is an interesting target for developing growth inhibitors for diverse Gram-positive bacterial strains because it is not present in eukaryotic cells. Researchers were motivated to create new pharmacophores because of restricted exposure to the LTA biosynthesis inhibitor. Drugs such as vancomycin and linezolid (1 and 2 mg/mL) were significantly less effective against MRSA than the new candidates of 1,3,4-oxadiazol-benzamides (ATCC 33592). According to the SAR, the activity is increased by highly electron-withdrawing groups like CF_3, F, and Cl, while it is decreased by hydrophilic moieties like OH, NH_2, tetrazole, and CN. The most effective analog with para-CF_3

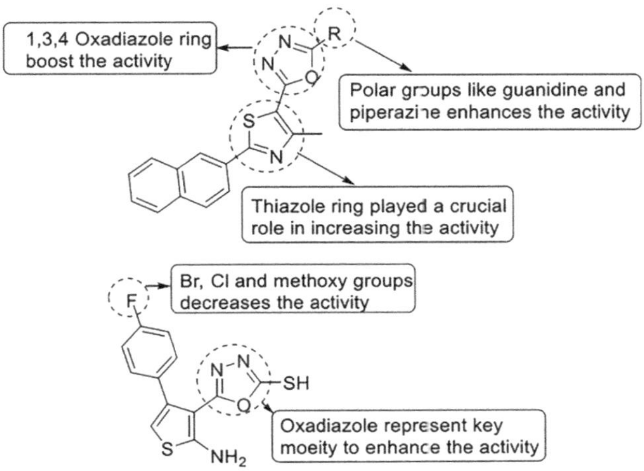

Figure 1. Oxadiazole compounds represent potent antibacterial agents against MRSA. (Verma et al. 2021a).

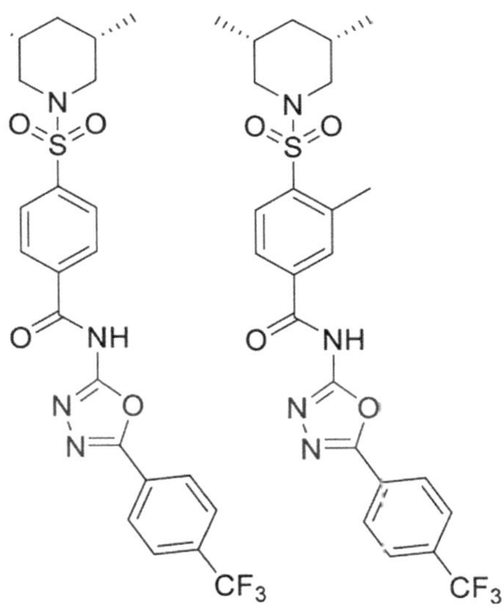

Figure 2. Selected benzamidooxadiazoles with improved potency against MRSA (Verma et al., 2021a).

(MIC: 0.25 mg/mL) and its meta-analog showed MIC of 11 mg/mL against MRSA, demonstrating the essential impact. Along with the common LTA biosynthesis inhibitor 1771, the molecule is chosen for the mechanistic action for its inhabiting bacterial infection (Naclerio et al. 2019) (Figure 2).

Due to low polarizability and high electronegativity, fluorine has attractive pharmacokinetic features that researchers have incorporated into molecular scaffolds to achieve excellent efficacy and bioavailability. For example, fluorine-containing groups such as trifluoromethylsulfonyl, pentafluorosulfonyl, trifluoromethylthio, and trifluoromethoxy play an excellent role when conjugated with oxadiazole. Most of these fluorine-containing groups are found in FDA-approved drugs and are essential for boosting the corresponding action. They are effective against MRSA, which exhibits highly tolerable to Caco-2 cells (colorectal cells), even at a higher dose of 128 mg/mL, which is 256 times more than the corresponding MICs against MRSA ATCC 33,592 and suggests these compounds to be a good antibiotic candidate. Furthermore, fluorine-decorated oxadiazoles were safe for RBCs at concentrations greater than 128 mg/mL and caused 50% hemolysis in RBCs, indicating only a low level of hemolysis in human RBCs. Combination therapy, such as radio and chemotherapy or combining two different drugs, is an emerging method of treating many infections (Naclerio et al. 2019, 2020).

The trans-translation represents a good target for antibiotics. Due to the participation in the ribosome rescue mechanism, in which transfer-messenger RNA (tmRNA)-SmpB is involved in proteolysis when used against *S. aureus*, a drug-resistant organism, small-molecule inhibitors like the 1,3,4-oxadiazole core have antibacterial properties. Results from biochemical analysis and cross-linking showed that it selectively inhibits bacteria from translating instead of doing so by binding to a location on 23S rRNA that is not recognized by any other drug. With MIC values of 0.64 mg/mL and 0.35 mg/mL against MRSA USA300 and 1.28 mg/mL and 0.7 mg/mL against Newman and MRSA252 strains, respectively, the derivatized drug inhibits numerous *S. aureus* strains, including MRSA. Following more research, it was utilized to compare it to antibiotics like daptomycin and chloramphenicol to determine the way of inhibiting *S. aureus*. While co-treatment with LL-37 (a human antimicrobial peptide) resulted in the synergistic growth inhibition of *S. aureus*, most antibiotics failed to demonstrate additive interaction and were, therefore, determined to be harmless to mammalian Hela cells and human HepG2 cells (Huang et al. 2019).

The difficult objective in the drug development program for antibacterial medicines is to create new pharmacophores with high biological half-lives and strong metabolic stability. An excellent place to start is with oxadiazole in the search for such antibacterial compounds. When tested against MRSA (2658 RCMB), structurally diverse thiazole conjugated 1,3,4-oxadiazoles showed inhibitory efficacy ranging from 4 to 16 mg/mL (Kotb et al. 2019). According to SAR, guanidine and hydrazine derivatives are produced when the side chain on nitrogen is extended, which results in a one-fold reduction in activity. Nitrogen-atom-containing cyclic rings further reduced the activity. However, it was discovered that for the eradication of MRSA biofilms, azithromycin was superior to vancomycin since it disrupted 41% of the adherent biofilm at 2X MIC concentration, but vancomycin only disrupted 34% of the MRSA biofilm mass at 32X MIC concentration. With a half-life of 5.5 hours, *in vivo* pharmacokinetic tests showed good metabolic stability. Compared to its guanidine analog, the oxadiazole ring significantly increased the bioavailability and stability (Verma et al. 2021b).

Alkyl/alkenyl/hydroxy alkenyl with varying chain lengths attached to the fifth position of oxadiazole with an aromatic and heteroaromatic ring at the second position has been developed in response to the antibacterial potential of oxadiazole analogs (Farshori et al. 2017). FtsZ is a bacterial division protein suggested as a potential target for synthesizing antibacterial drugs to stop the growth of drug-resistant bacteria. According to the study, a methicillin-resistant strain of *S. aureus* (ATCC43300) has a MIC of 1 mg/mL, which is equal to a penicillin (MIC: 1 mg/mL), superior to linezolid (MIC: 2 mg/mL) and erythromycin (MIC: > 64 mg/mL), and inferior to ciprofloxacin (MIC: 0.25 mg/mL). According to the SAR analysis, the longer alkyl chain facilitates switching the positions of the substituents on the oxadiazole ring and replacing the oxadiazole with a heterocycle with negative activity (Bi et al. 2019).

Oxadiazole's bioavailability needed to be improved due to its extensive use in medicines as antibiotics. This included enhancing efficacy, decreasing cytotoxicity, raising selectivity, and creating new oxadiazole hybrids to combat rapidly emerging drug-resistant bacterial strains. Since many bacteria are now resistant to norfloxacin, a derivatized form of the drug must be created in order to combat bacterial resistance. In order to increase the hydrophobicity and the activity of norfloxacin, 1,3,4-oxadiazole has been added to its structural design. Three clinical isolates of MRSA1-3 were tested against four norfloxacin-1,3,4 oxadiazole analogs. Unexpectedly, the compounds demonstrated superior action to the precursor norfloxacin, with MICs in the range of 0.5, 0.25, and 1 mg/mL.

According to the SAR analysis, methoxy, chloro, bromo, and methyl groups were appropriately substituted, which increased the activity that efficiently damaged the membrane of MRSA2 and caused irreparable harm to the bacteria (Guo et al. 2019). A large variety of biologically active scaffolds rely on quaternary ammonium salts (QAS), which typically adsorb on negatively charged species on cell walls and impact the cells. As a result, 1,3,4-oxadiazoles were also used to make quaternary salts to test their antibacterial effectiveness against MRSA. The inhibitory results indicate that synthetic quaternary salts of oxadiazoles were 100 times more effective against MRSA than the control gentamicin (MIC: 100 mg/mL) (ATCC 43300). Based on the findings, 1,3,4-oxadiazole, and QA's derivatives could make an attractive model platform for studies of antibacterial drugs in the future (Rohand et al. 2019).

In another study, dihydropyrimidine bound to 1,3,4-oxadiazole analogs was tested against five different bacterial strains, including MRSA (ATCC 43300), which demonstrated moderate to better activity and did not exhibit much toxicity to HEK293 (human embryonic kidney cells) and less hemolytic activity in human whole blood cells (Dinesh and Jignasu 2019). It is well known that biofilms exhibit strong antibiotic resistance, although, in the presence of antimicrobials, biofilms could not perform as well as planktonic cell development (Spoering and Lewis 2001). It is advised to create newer antibacterial agents that can successfully prevent biofilm development and eliminate planktonic cells' capacity for resistance. Three novel 1,3,4-oxadiazole derivatives were tested to see how well they worked against *S. aureus* biofilms and planktonic cells as antibacterial agents. When tested against two MRSA strains with MBC values ranging from 8 to 64 mg/mL, all three drugs demonstrated inclusive inhibition. *S. aureus* MW2 and USA300 planktonic cells were utilized to test the antimicrobial activity at 4 X MIC concentration, and all the compounds were toxic to *S. aureus* growth. Additionally, it demonstrated biofilm inhibition at a dosage of 16 mg/mL, which is sufficient to stop the growth of *S. aureus* and biofilms (Zheng et al. 2018).

The α-amino phosphate group's demanding biological activity profile drew researchers to include it in a bioactive heterocyclic system to increase the effectiveness of the molecular scaffolds against various biological targets. A unique group of 1,3,4-oxadiazole derivatives that contain α-amino phosphate has been combined and tested for their biological effects on a variety of bacterial species, including MRSA. Compared to comparable amide analogs and the positive control drug gentamicin, α-amino phosphate derivatives of oxadiazole efficiently suppressed the growth of the MRSA strain with a zone of inhibition spanning 17 to 29 mm. According to SAR, a key factor in modifying the activity was the location of EDG and EWGs on the aromatic ring next to the amino phosphate moiety. While mono substitution of fluoro and bromo resulted in lower activity, and strong electron-withdrawing (NO_2) and strong electron-donating (OMe) groups weaken the activity, di-substitution of EWGs on the aromatic ring of a-amino phosphate derivatives (Cl and F) augment the activity (Boshta et al. 2018). Phenylthiazoles have been found to be effective antibiotics against multidrug-resistant staphylococcal strains. The effectiveness of each molecule was tested against a variety of bacterial infections, including MRSA. It showed inhibition ranges from 12.5 to 50 mg/mL, which is better than the MIC of ciprofloxacin (MIC: 6.25 mg/mL) (Mohammad et al. 2014; Seleem et al. 2016). By adding a nitrogenous head and a lipophilic tail, these compounds can be made more metabolically stable, although doing so degrades their water solubility and makes their aqueous solubility poorer (Mohammad et al. 2015). The low oral bioavailability of antibacterial drugs with poor solubility is a key problem for all antibiotics. A polar oxadiazole ring with fewer carbon atoms and a higher polar surface area has been used to solve the problem of oral bioavailability. Secondary and tertiary amines on side chains of thiazole-conjugated oxadiazoles with nitrogenous heads were ineffective against MRSA (2658 RCMB). While nitrogenous head such as guanidine were discovered to be equally powerful as vancomycin, nitrogenous head such as hydrazine, piperazine, and hydroxyl azetidine showed superior MIC values. Increased morpholine and piperazine ring polarity on the nitrogenous sidechain further boost their anti-MRSA efficacy, with MICs of 1.56 and 1.17 mg/mL, respectively (2658 RCMB). Compared to vancomycin, the lead compound's time-kill assay against MRSA indicated it is bactericidal. *In vivo* PK profiling revealed that the molecule had an oral bioavailability of 42.5%, was non-toxic at 512 mg/mL to human colorectal cells (HRT-18), and had good aqueous solubility (Saq 14 104 mM) (Hagras et al. 2018)

In order to test the ability to suppress MRSA biofilms, bromopyrrole alkaloids were derivatized with 1,3,4-oxadiazole, which has a thio-alkyl/aryl/heteroaryl ring in the second position of the ring (Rane et al. 2015). Most of the derivatives demonstrated significant biofilm inhibition against MRSA; the most potent compound was fourfold more active (MBIC: 0.78 mg/mL) than vancomycin (MBIC: 6.25 mg/mL), while the remainder of the molecules were equivalent to vancomycin. While thiophenyl and thio-ethyl at the second position on oxadiazole diminish the ant-biofilm activity, SAR suggests that halogen-, amino-, nitro-, and methoxy groups at the fourth position on the phenyl ring were superior for the biofilm inhibition. The substance's antivirulence property is indicated by the fact that it did not affect the viability of bacteria at antibiofilm activity concentration (Verma et al. 2021a). Ribonucleotide N5-carboxy-amino-imidazole (N5-CAIR) Mutase (PurE) is a bacterial enzyme that produces 4-carboxy-amino-imidazole ribonucleotide (CAIR) from 5-amino-imidazole ribonucleotide (AIR) and carbon dioxide (CO_2) (Zhang et al. 2008; Tranchimand et al. 2011). It has been acknowledged that the *de novo* purine biosynthesis route is a legitimate target for discovering antibacterial compounds. The most effective method for finding hit compounds that can bind to a particular target in a biological system is high-throughput screening (HTS).

A new class of oxadiazole-based phenyl-thiolatobismuth (III) complexes have been tested for their ability to combat MRSA. Compared to the simple oxadiazole thiones, the complex mono-phenyl bismuth (III) analogs demonstrated a wider range of action. With a MIC value of 1.1 mM, which is on par with vancomycin, complex [BiPh(Me-PTOT)$_2$], it demonstrated dominating action against MRSA. Changing the para-substituents of the phenyl ring in the oxadiazole ring did not affect the activity profile. These compounds' nontoxicity to cultivated mammalian COS-7 cell lines, even at 10 X MIC values (20 mg/mL), is a strong argument in favor of them. These complexes provide new opportunities for identifying antibacterial drugs due to their great proficiency and low toxicity (Kim et al. 2015; Luqman et al. 2015).

In order to test the antibacterial effectiveness against MRSA, 4-thiazolidinedione has been attached to an oxadiazole ring while investigating various heterocycle conjugated derivatives (Narasimhamurthy et al. 2014; Girish et al. 2014a). Two new chemical libraries have been created and tested against two MRSA clinical isolates (CCARM 3167 and CCARM 3506) exhibit MIC ranges from 1 to 64 mg/mL. The compound that was chosen as the best in the library showed a MIC of 1 mg/mL against both MRSA strains, and it is equivalent to moxifloxacin (MIC: 1 mg/mL), two times more effective than gatifloxacin (MIC: 21 mg/mL), 64 times more active than oxacillin (MIC: > 64 mg/mL), and 4 to 8 times better than norflox. The thiazolidinone ring on the oxadiazole moiety could be further modified to improve the antibacterial action (Liu et al. 2014). Through the intramolecular cyclization of 2-(2-aroylaryloxy) acetohydrazides, ten 2-[2-(aroyl)aroyloxy]methyl-1,3,4-oxadiazole derivatives had been synthesized. These novel, structurally unique oxadiazole moieties were tested against MRSA and other bacterial strains and showed excellent inhibition activity (Girish et al. 2014b). Numerous naturally occurring compounds with nitropyrrole bases showed effective antibacterial properties (Wang et al. 1975; Santo et al. 1998; Raju et al. 2010; Kwon et al. 2010). The ability of 4-nitropyrrole-based chloramphenicol to combat drug-resistant *Staphylococcus aureus* germs with MIC values up to 0.8 mg/mL was also explained in a few papers. Given the effectiveness of nitro pyrrols, an effort was made to create 4-nitro pyrrole derivatives that contained 1,3,4-oxadiazole to increase the efficacy further against drug-resistant bacteria, particularly MRSA (Rane et al. 2013). Due to the presence of many methyl groups, the development of 1,3,4-oxadiazol-benzamide analogs as antibacterial agents against methicillin- and vancomycin-resistant bacteria has proven particularly useful. At a MIC of 2 mg/ml, it prevents bacterial growth. The antibacteral efficacy against MRSA *in vivo* was proven via the skin wound infection model (Opoku-Temeng et al. 2018). Living organisms are seriously threatened by drug-resistant bacterial infections. Multidrug-resistant bacteria are extremely difficult to treat. Oxadiazole will be a powerful inhibitor against bacterial infection thanks to its structural diversity, which promotes the development of many derivatives and thus demonstrates moderate to good activity against various bacterial strains.

2.2 Inhibitors Against MRSA as Antibacterial Agents

To better understand the SAR, 72 structurally diverse indole and pyrazole-substituted oxadiazole analogs were tested *in vitro* and *in vivo* for their antibacterial efficacy against ESKAPE strains, with a particular focus on *S. aureus* (ATCC 29213) (Boudreau et al. 2020). Methoxy and trifluoromethoxy at para-position and azide group at any position of the ring showed substantial MIC values against *S. aureus*. Hydrophobic substituents, especially halogens (F, Cl, Br, and I), on diphenyl ether rings at different places also worked well against *S. aureus*. Other substituents with low activity include methyl, alkyl carbamates, amides, tert-butyl, amines, and sulfonates. A phenyl ring's activity against *S. aureus* is generally reduced by hydrogen-bond donor groups, and heterocyclic moieties like benzothiazole and pyridine can replace a phenyl ring and still be somewhat effective. Tetrahydrofuran, pyrrolidines, azetidines, and other hetero aliphatic ring systems fared inadequately against *S. aureus* (Boudreau et al. 2020) (Figure 3).

In the mouse peritonitis model of MRSA infection, oxadiazoles constitute a non-β-lactam inhibitor of cell-wall biosynthesis with a good action profile against multidrug-resistant MRSA and low clearance, more half-life and oral bioavailability of 97% (Janardhanan et al. 2016). A synergistic study with β-lactam and non-β-lactam antibiotics reveals the beneficial synergistic effect of β-lactam antibiotics. Combination therapy is a useful strategy to battle drug-resistant bacterial strains. Oxacillin's *in vivo* efficacy in a mouse neutropenic thigh infection model proved that combination therapy is more effective than single-agent therapy (Ceballos et al. 2018).

The best treatment for both community- and hospital-acquired infections was an oxyazolidinone-based linezolid antibiotic drug with a distinct mechanism of action, dynamic pharmacokinetic properties, and 100% oral bioavailability, but soon after its introduction, it developed resistance to numerous multidrug-resistant bacteria (Dryden et al. 2011). Since 1,2,4-oxadiazole is a superior substitute for oxazolidinone, Fortuna et al. 2014 development of new antibacterial to combat drug-resistant microorganisms also included thoroughly examining the SAR. A substantial amount of activity was seen with one of the series of drugs tested against clinical isolates of MRSA that were multidrug-resistant, with MIC values comparable to those of linezolid (1.6 mg/mL) at 3.125 mg/mL. In contrast to the aromatic fluoro substitution, the SAR demonstrated the additive ability of triazole to improve activity in the oxadiazole side chain. The racemic mixing state was blamed for the compounds' lower action (Fortuna et al. 2014).

Through the synthesis of a large library of 59 derivatives of structurally various oxadiazole derivatives, the SAR of 1,2,4-oxadiazoles was investigated to create

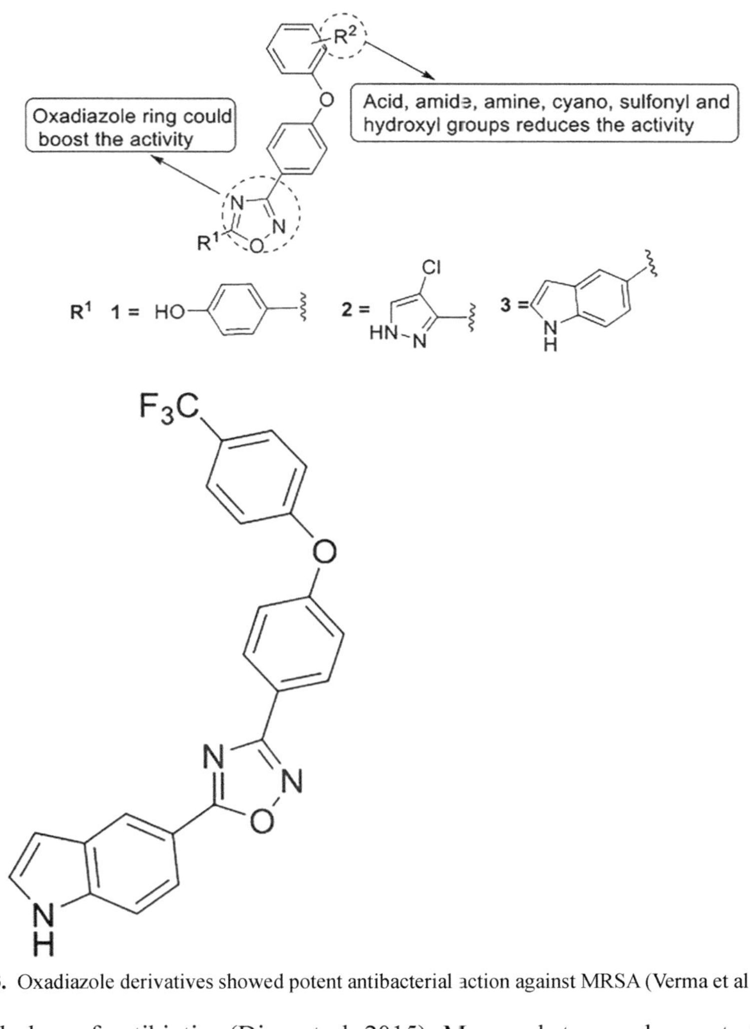

Figure 3. Oxadiazole derivatives showed potent antibacterial action against MRSA (Verma et al. 2021a).

a novel class of antibiotics (Ding et al. 2015). Many substances demonstrated the necessary action against *S. aureus* and numerous MRSA strains. According to the SAR, an H-bond donor in the oxadiazole's phenyl ring at position C-5 is necessary for antibacterial activity. Pyrazole, aniline, and phenol all effectively inhibit MRSA through hydrogen bonding. Meanwhile, hydrogen-bond acceptors on the phenyl ring at position C-5 of the oxadiazole ring and carboxylic acids slow the antibacterial action. Some aliphatic heterocycles (piperidine and piperazine) and heterocyclic systems like pyrroles, pyrazoles, and triazoles lose their activity. Considering their preparation, the top 100 compounds were selected from the *in-silico* analyses of oxadiazole-based derivatives. When they were all tested against the ESKAPE family of bacteria, hit molecule 42 (MIC: 2 mg/mL against *S. aureus*) was discovered and is now the subject of additional research. Macromolecular synthesis experiments indicated that PBP2a is inhibited by blocking cell-wall biosynthesis rather than replication, transcription, or translation (penicillin-binding proteins). The halogens and hydrogen-bond acceptor moieties on the phenyl ring at the C-5 of oxadiazoles decrease their activity, as indicated by SAR studies. The *in vitro* action is enhanced by pyrazole ring substitution, but mammalian cells are poisoned. The activity was not significantly affected when sulfur and nitrogen atoms were used in place of the ether oxygen linker. The investigated substances showed up to 100% oral bioavailability, a half-life of up to 40 hours, and were proven to be bactericidal. Animal models for neutropenic thigh infection and murine peritonitis/sepsis were used to test the *in vivo* effectiveness of these lead candidates. In linezolid-susceptible strains and animal models infected with linezolid-resistant strains, compound 39 had equivalent efficacy to linezolid (Verma et al. 2021a). A variety of structurally modified derivatives of linezolid with carboxamide side chains and varying numbers of fluorine atoms on the phenyl ring at the C-5 on the oxadiazole ring were synthesized using linezolid decorated with a 1,2,4-oxadiazole moiety in the place of its isoster oxazolidinone ring (Palumbo Piccionello et al. 2012). All of the produced compounds were tested against MRSA and other bacterial strains. Sadly, none of the substances were effective against MRSA (C530).

In order to create new and effective antibacterial medicines that can combat MRSA, a different series of linezolid-like oxadiazole derivatives has been rationalized (Fortuna et al. 2013). The substitution of 1,2,4-oxadiazole for morpholine and the insertion of various groups to the side chains of oxazolidinone was responsible for the shift in the *in vitro* antibacterial activity of these novel families of antibiotics. Four of the 14 studied compounds demonstrated action against MRSA that was comparable to or better than that of linezolid, whereas sulfur-embedded variants outperformed regular linezolid in this regard. Linezolid outperformed two acetamide analogs against the MRSA strain, presumably due to the racemic mixing state. The efficacy of these kinds of substances against MRSA may be significantly enhanced by their pure enantiomers. Cell survival tests on the PK15, HaCaT, and HepG2 cell lines demonstrated the decreased toxicity of the linezolid-like compound due to the impact of the fluorine atom.

In contrast, 1,2,4-triazole and 1,3-diazole with 5-acetamido methyl groups resulted in a complete loss of activity against MRSA, according to the SAR, but other heterocycles like triazole, pyridine, and isoxazole substituents were able to preserve or augment the activity. The activity is drastically reduced by an eightfold increase in their MIC values when the aceto amido methyl substituent is replaced with a thio aceto amido methyl group. The effectiveness of the chemical has been further supported by modeling results that dock the S-enantiomer in complex with *S. aureus* over linezolid in complex with *Deinococcus radiodurans* (Verma et al. 2021a). The 1,2,4-oxadiazole derivatives with chromone substitutions were synthesized and tested for their *in vitro* antibacterial efficacy against several bacterial strains, including MRSA. All bacterial strains displayed poor activity profiles for mono-substituted (-Cl, -Br, eF, and eCH_3) and unsubstituted chromones. While the activity was marginally improved by the di-substitution of chloro and methyl groups on chromone ring 45f, it was still inferior to that of ordinary gentamicin and superior to that of erythromycin, ampicillin, and ciprofloxacin (Diwakar et al. 2011).

Liu and coworkers have integrated pyrimidine fused oxadiazole derivatives and tested their antibacterial effectiveness against several MRSA strains to study the biological potential of various heterocycles (Verma et al. 2021a). The SAR indicated that bromo-substituted 5H-[1,2,4]oxadiazolo[4,5-a]pyrimidine analogs fared better against MRSA than the equivalent fluoropyrimidine analogs, indicating that they are a promising pharmacophore for anti-MRSA drugs. The chemical 46 g was discovered to be more effective in the library, exhibiting superior activity to vancomycin and cefepime against MRSA 128 and MRSA 747, as well as three times the activity of normal cefepime against the MRSA 42 strain. All of the strains employed in the study exhibited more activity when dihalogen substitution was used as a pattern; interestingly, all compounds demonstrated stronger potency against Gram-positive bacterial strains than Gram-negative bacterial strains (Liu et al. 2018).

3. Role of Histone Deacetylase (HDAC) Inhibitors in Anticancer Therapy

Commercially available anti-cancerous drugs suffer from drawbacks such as decreased effectiveness, drug resistance, unfavorable side effects, and non-selectivity. This has generated interest in developing a newer, more potent inhibitor of cancer with fewer side effects urgently necessary. Recently, several drugs such as Vorinostat (SAHA), Romidepsin (FK-228), Beli nostat (PXD-101), Panobinostat (LBH-589), and Chidamide have been licensed by US/Chinese FDA for cancer treatment and several are in the queue of clinical trials. Since many years ago, heterocycles oxadiazole and their derivatives have been extensively used as inhibitors of cancer cells, which has attracted a lot of attention among the research community. Inhibiting various growth factors and enzymes, such as histone deacetylase (HDAC), thymidylate synthase (TS), vascular endothelial growth factor (VEGF), glycogen synthase kinase-3 (GSK), epidermal growth factor (EGF), and others, is a key component of the important mechanism of suppressing cancer cells (Matore et al. 2022). Currently, HDAC is seen as a possible therapeutic target. Apoptosis, differentiation, cell cycle arrest, suppression of DNA repair, up-regulation or reactivation of silent tumor suppressors, and down-regulation of growth factors are some of the different pathways that HDAC inhibitors control and induce cancer cell death. Hence, oxadiazole and its derivatives are widely recognized as novel anticancer drugs because they inhibit HDAC enzymes (Matore et al. 2022).

Recently, tri fluoro acetyl thiophene carboxamides were studied as HDAC inhibitors with moderate selectivity for the class II HDAC enzymes. They discovered that numerous derivatives exhibited strong activity against HDAC4 after substituting various bioisosteric heteroaromatic oxadiazoles and thiazoles for the carboxamide moiety. It demonstrated greater selectivity for HDAC1 and HDAC6 by 685 and 75 times, respectively (Muraglia et al. 2008). HDAC inhibitors with a scaffold moiety of 1,2,4-oxadiazole have assessed *in vitro* anticancer activities. Nine of them demonstrated the strongest histone deacetylase (HDAC) inhibitory action, with IC_{50} values that were lower than SAHA, against HDAC1, HDAC2, and HDAC3. The structural characteristics were demonstrated using a 3, 5-disubstituted 1, 2, 4-oxadiazole as the linker moiety. *In vitro*, anti-proliferation assay results indicated compounds were more effective than regular SAHA. The acetylating process is greatly upregulated based on western blot assays, and the mechanism of action was also disclosed by molecular docking research. The mechanism was investigated by a flow-cytometry study, and it operates through apoptosis and produces cell cycle arrest (Yang et al. 2019a). As novel 1,3,4 oxadiazole with glycine/alanine hybrids as HDAC8 specific inhibitors, the preliminary evaluation study revealed that the 1,3,4 oxadiazole with alanine hybrid (R)-2-amino-N-((5-phenyl-1,3,4-oxadiazol-2-yl) methyl) propenamide. Additionally, it demonstrated a dose-dependent impact and a drop in the percentage of apoptotic cells and mitochondrial membrane potential, as well as lower IC_{50} values against the MDA-MB231 and MCF7 breast cancer cell lines (Pidugu et al. 2017).

Different pharmacophoric groups can be substituted in the oxadiazole ring to create novel compounds with improved anticancer potential. The development of Hydrox-mate-based HDAC inhibitors with a 1,2,4-oxadiazole ring moiety was investigated. The investigation of the US FDA-approved drug Vorinostat is the basis for this study (SAHA). Only the cap group of Vorinostat has had various groups substituted and examined for anticancer profiles that were more powerful than conventional Vorinostat. SAR generally illustrates changes in compound activity brought on by substitutions made to the Vorinostat compound's fundamental structural elements. According to the study, the type of the numerous substituents rather than the oxadiazole's fundamental structure determines its activity. The series demonstrates that the 1,3,4-oxa diazole's fifth position substitute greatly affects its anticancer potential. The 1,3,4-oxadiazole ring demonstrates increased inhibitory activity as its electronegativity rises. Additionally, Ortho > Meta > Para is the order of substitution with increasing activity. The order of substitution for halogens was I > Br > Cl > F in the ortho position of the benzene ring, while Br > Cl and Br > Cl > I > F were the orders in the meta and para positions, respectively (Yang et al. 2019b). Biologically active compounds contain small oxadiazole nuclei showing excellent anticancer properties. Several 1,3,4-oxadiazole derivatives were tested for their ability to inhibit various HDAC enzyme types. The most effective and selective HDAC1 inhibitor was (E)-N-(2-aminophenyl)-3-(4-(((5-(naphthalene-1-yl methyl)-1,3,4-oxadiazol-2-yl) methyl) phenyl) acrylamide (4). Comparatively, to SAHA, it prominently inhibited HDAC1. It also exhibits a strong antiproliferative effect against the five leukemia cancer cells and its HDAC inhibitory action against HDAC1/4/6. The benzamide derivatives were less effective in inhibiting HDAC4/6 than the hydroxamic acid derivatives. Whether doxorubicin was used with HDAC inhibitors was also studied. It was found that doxorubicin plus HDAC inhibitor combinations showed greater cell growth arrest. The combination of HDAC inhibitors with various chemotherapeutic medicines (phosphatase inhibitors, GSK3b inhibitors, etc.) has demonstrated excellent synergy due to the multi-factorial molecular aspect of cancer (Valente et al. 2014).

A new 2,5-disubstituted 1,3,4-oxadiazoles were synthesized to examine their anticancer potential by HDAC inhibitory assay and antiproliferative assay. By specifically inhibiting HDAC8, all the evaluated oxadiazole compounds demonstrated significant anticancer efficacy. Comparable to the positive control medication Vorinostat, the compound demonstrated significant HDAC8 inhibitory action as well as improved anticancer efficacy. Additionally, it demonstrated nM-range HDAC1 and HDAC2 inhibitory action, which effectively inhibited the growth of MDA MB231 breast cancer cell lines. The most potent hybrid has been synthesized by replacing the aromatic phenyl ring at position two and the alanine framework at position five of the 1,3,4-oxadiazole scaffold, which eventually inhibits cancerous activity (Pidugu et al. 2016).

Three series of oxadiazole analogs, including hydroxamate, 2-aminobenzamide, and trifluoromethyl ketone, were developed, and their MTT-based assay for *in vitro* antiproliferative efficacy was tested. Further, the substances were evaluated against HDAC1, 2, and 8 isoforms. The investigations suggest these compounds might be promising for developing antitumor medicines with HDAC inhibitory action. The majority of the 2-aminobezamide and hydroxamate analogs showed better selectivity for HDAC1 than for HDAC2 and HDAC8. Compared to the controls, Vorinostat and Entinostat demonstrated the strongest inhibition with lower IC_{50} values. Against all evaluated HDAC enzymes and cancer cell lines, all three of these drugs demonstrated strong anticancer activity in the M range. These three compounds comprise substituted 1,2,4-oxadiazole, six carbon alkyl chains, and hydroxamic acid (ZBG) (CAP). Significant HDAC1 inhibitory activity was seen when 4-nitro phenyl (15), thiophene (16), and pyridine (17) were substituted (Cai et al. 2015a).

A number of dual inhibitors focusing on HDAC6 and Hsp90 had been synthesized. In order to find effective HDAC6 and Hsp90 inhibitors, the first choice was ZINC and ChEMBL open databases, followed by ADMET screening of the entire database and molecular docking. The ten best hits were selected from the molecular docking results and used in the experiments. The 1,2,4-oxadiazole ring served as the cap structure for all ten potent hits, which was an interesting discovery. Various inhibitory tests employing SAHA, Tubastatin (TUB), and 17-AAG were used to investigate the inhibitory activity (% inhibition) against HDAC1, HDAC6, and Hsp90. According to the HDAC6 inhibitory assay results, seven substances exhibit more than 50% inhibition at a concentration of 50 μM. The two compounds appeared to exhibit potent inhibition when these compounds were studied further at a concentration of 5 μM. A 1,2,4-oxadiazole derivative with para substitution was more effective than one with meta substitution. HDAC1 and Hsp90 have not been effectively inhibited by a single substance (Pinzi et al. 2020). By altering the CAP structure with a -carboline 1,3,4-oxadiazole hybrid, various new HDAC inhibitors have been synthesized. The antitumor activity of the produced compounds was assessed utilizing the HDAC inhibitory and MTT tests. Five distinct HDAC enzymes—HDAC1, 2, 3, 4, and 8—and the cancer cell line HCT116 were used to examine each substance. Compared to normal SAHA, it demonstrated a reduced IC_{50} against HDAC1, 2, and 6. Additionally, the MTT assay demonstrated its potency by demonstrating maximal inhibition on HCT116 cancer cell lines with a lower IC_{50} than typical SAHA. Additionally, docking investigation also demonstrated that high binding affinity displayed numerous favorable interactions with certain amino acid residues of HDAC protein. The modified CAP structure (-carboline 1,3,4-oxadiazole) of 24 demonstrated improved activity and three positive interactions with amino acid residues (Tian et al. 2022).

Using a variety of 2-amino benzamide and hydroxamate derivatives with an oxadiazole nucleus, the *in vitro* anticancer properties were examined using an MTT-based assay on the human cell lines U937, A549, NCI-H661, MDA-MB-231, and HCT116. The 2-aminobenzamide class of substances exhibited the greatest inhibition against the U937 cell line for human acute monocytic myeloid leukemia. Additionally, the derivatives were examined for their capacity to inhibit HDAC1, 2, and 8. Compared to HDAC2 and HDAC8, all 2-Amino benzamide compounds demonstrated a better selectivity for HDAC1. Compared to SAHA and entinostat, compounds displayed greater potency and lower IC_{50} values. All compounds have the cap structure 1,2,4-oxadiazole; however, they differ in the third position replacement. The substitutions comprise 4-nitro phenyl, 2-chloro phenyl, 4-chloro phenyl, and 4-fluro phenyl at the third position of the 1,2,4-oxadiazole cap structure. Additionally, these substances demonstrated strong anticancer activity against five distinct cancer cell lines. In contrast to the compounds in the 2-aminobenzamide series, most hydroxamate derivatives selectively inhibit HDAC8 over HDAC1 and 2 (Cai et al. 2015b).

Using hydroxamic acid, two novel series of 2-[5-(4-substituted phenyl)-[1,3,4]-oxadiazol/thiadiazol-2-ylamino]-pyrimidine-5-carboxylic acid (tetrahydro-pyran-2-yloxy)-amides were synthesized and tested for their histone deacetylase inhibitory efficacy. The findings of *in vitro* anticancer tests' demonstrated that compounds were most effective against HDAC1 and HCT-116. These substances have the strongest CAP structures, which are 1,3,4-oxadiazole scaffolds. The second position of the oxadiazole underwent various substitutions, whereas the substitution at the fifth position was the same in all four compounds (pyrimidine as a linker and hydroxamic acid as ZBG). Similar activity against HDAC1 was displayed by the compounds which were replaced with aromatic phenyl and 3-chloro phenyl. This information revealed that the 1,3,4-oxadiazole's substituted phenyl ring at position two is necessary for activity and that other substitutions at different phenyl ring positions can change or modify the HDAC inhibitory activity. These findings can be applied to developing new HDAC inhibitors (Rajak et al. 2011).

4. Conclusion

Oxadiazole-based hybrids demonstrated great biological significance and have been effective as MRSA growth inhibitors and anti-HDAC activity for cancer treatment. A novel heterocyclic small-molecule-based MRSA inhibitor, which is supported by an array of oxadiazole derivatives to combat MRSA in both *in vitro* and in vivo conditions, is now being heavily researched. Their antibacterial action against MRSA displayed various modes of action. Adding different substituents with different electronic properties to some aryl/heteroaryl moieties fixed to the parent oxadiazole skeleton significantly impacted whether the antibacterial activity against MRSA was increased or decreased. On the other hand, angiogenesis, which is controlled by HDAC and the release of VEGF, is one of the most significant processes. It is ultimately responsible for the development of cancer and metastasis. HDAC, an authentic and potential target enzyme, is widely considered for developing anticancer drugs. In order to achieve the appropriate potency, placements of substituents inside the oxadiazole moiety with varied pharmacophores are highly important. This supported the significance of oxadiazole scaffolds in the development of anticancer medications. Researchers across the globe should put more focus on and learn the fundamental requirements and critical elements for developing newer HDAC inhibitors with minimal adverse effects, as well as MRSA growth inhibitors.

References

Bassetti, M., G. Poulakou, E. Ruppe, E. Bouza, S. J. V. Hal and A. Brink. 2017. Antimicrobial resistance in the next 30 years, humankind, bugs and drugs: a visionary approach. Intensive Care Med. 43: 1464–1475. https://doi.org/10.1007/s00134-017-4878-x.

Bi, F., D. Song, Y. Qin, X. Liu, Y. Teng, N. Zhang et al. 2019. Discovery of 1,3,4-oxadiazol-2-one-containing benzamide derivatives targeting FtsZ as highly potent agents of killing a variety of MDR bacteria strains. Bioorg. Med. Chem. 27(4): 3179–3193.

Bi, G. and G. Jiang. 2006. The molecular mechanism of HDAC inhibitors in anticancer effects. Cell. Mol. Immunol. 3(4): 285–290.

Boshta, N. M., E. A. Elgamal and I. E. T. El-Sayed. 2018. Bioactive amide and a-aminophosphonate inhibitors for methicillin-resistant Staphylococcus aureus (MRSA). Monatsh. Chem. 149: 2349–2358.

Boudreau, M. A., D. Ding, J. E. Meisel, J. Janardhanan, E. Spink, Z. Peng et al. 2020. Structure activity relationship for the oxadiazole class of antibacterials. ACS Med. Chem. Lett. 11(3): 322–326.

Cai, J., H. Wei, K. H. Hong, X. Wu, M. Cao, X. Zong et al. 2015a. Discovery and preliminary evaluation of 2-aminobenzamide and hydroxamate derivatives containing 1,2,4-oxadiazole moiety as potent histone deacetylase inhibitors. Eur. J. Med. Chem. 96: 1–13.

Cai, J., H. Wei, K. H. Hong, X. Wu, X. Zong, M. Cao et al. 2015b. Discovery, bioactivity and docking simulation of Vorinostat analogues containing 1,2,4-oxadiazole moiety as potent histone deacetylase inhibitors and antitumor agents. Bioorg. Med. Chem. 23(13): 3457–3471.

Carmeliet, P. 2005. VEGF as a key mediator of angiogenesis in cancer, Oncology 69(3): 4–10.

Ceballos, S., C. Kim, D. Ding, S. Mobashery, M. Chang and C. Torres. 2018. Activities of oxadiazole antibacterials against *Staphylococcus aureus* and other Gram-positive bacteria. Antimicrob. Agents Chemother. 62(8): 00453–00418.

Chen, X., J. Leng, K. P. Rakesh, N. Darshini, T. Shubhavathi, H. K. Vivek et al. 2017. Synthesis and molecular docking studies of xanthone attached amino acids as potential antimicrobial and anti-inflammatory agents. Med. Chem. Comm. 8(8): 1706–1719.

Dinesh, B. M. V. R. G. and P. M. Jignasu. 2019. Cyclization and antimicrobial evolution of 1,3,4-oxadiazoles by carbohydrazide. World Sci. News 124(2): 304–311

Ding, D., M. A. Boudreau, E. Leemans, E. Spink, T. Yamaguchi, S. A. Testero et al. 2015. Exploration of the structure activity relationship of 1,2,4-oxadiazole antibiotics. Bioorg. Med. Chem. Lett. 25(21): 4854–4857.

Diwakar, S. D., R. S. Joshi and C. H. Gill. 2011. Synthesis and *in vitro* antibacterial assessment of novel chromones featuring 1,2,4-oxadiazole. J. Heterocycl. Chem. 48(4): 882–887.

Dryden, M. S. 2011. Linezolid pharmacokinetics and pharmacodynamics in clinical treatment. J. Antimicrob. Chemother. 66(4): iv7eiv15.

Eckschlager, T., J. Plch, M. Stiborova and J. Hrabeta. 2017. Histone deacetylase inhibitors as anticancer drugs. Int. J. Mol. Sci. 18(7): 1414.

Farshori, N. N., A. Rauf, M. A. Siddiqui, E. S. Al-Sheddi and M. M. Al-Oqail. 2017. A facile one-pot synthesis of novel 2,5-disubstituted-1,3,4-oxadiazoles under conventional and microwave conditions and evaluation of their *in vitro* antimicrobial activities. Arab. J. Chem. 10(2): S2853–S2861.

Ford, C. W., J. C. Hamel, D. M. Wilson, J. K. Moerman, D. Stapert, R. J. Yancey et al. 1996. *In vivo* activities of U-100592 and U-100766, novel oxazolidinone antimicrobial agents, against experimental bacterial infections. Antimicrob. Agents Chemother. 40(6): 1508.

Fortuna, C. G., C. Bonaccorso, A. Bulbarelli, G. Caltabiano, L. Rizzi, L. Goracci et al. 2013. New linezolid-like 1,2,4-oxadiazoles active against Gram-positive multiresistant pathogens. Eur. J. Med. Chem. 65: 533–545.

Fortuna, C. G., R. Berardozzi, C. Bonaccorso, G. Caltabiano, L. Di Bari, L. Goracci et al. 2014. New potent antibacterials against Gram-positive multiresistant pathogens: effects of side chain modification and chirality in linezolid-like 1,2,4-oxadiazoles. Bioorg. Med. Chem. 22(24): 6814–6825.

Frimodt-Meller, N., M. W. Bentzen and V. F. Thomsen. 1986. Experimental infection with Streptococcus pneumoniae in mice: correlation of *in vitro* activity and pharmacokinetic parameters with *in vivo* effect for 14 cephalosporins. J. Infect. Dis. 154(3): 511–517.

Girish, V., N. F. Khanum, H. D. Gurupadaswamy and S. A. Khanum. 2014a. Synthesis and evaluation of *in vitro* antimicrobial activity of novel 2-[2-(aroyl)aroyloxy] methyl-1,3,4-oxadiazoles. Russ. J. Bioorg. Chem. 40(3): 330–335.

Girish, Y. R. K. S. Sharath Kumar, U. Muddegowda, N. K. Lokanath, K. S. Rangappa and S. Shashikanth. 2014b. ZrO2-supported Cu(ii)eb-cyclodextrin complex: construction of 2,4,5-trisubstituted-1,2,3-triazoles via azideechalcone oxidative cycloaddition and post-triazole alkylation. RSC Adv. 4(99): 55800–55806.

Guo, Y., T. Xu, C. Bao, Z. Liu, J. Fan, R. Yang et al. 2019. Design and synthesis of new norfloxacin-1,3,4-oxadiazole hybrids as antibacterial agents against methicillin-resistant *Staphylococcus aureus* (MRSA). Eur. J. Pharmaceut. Sci. 136: 104966.

Hagras, M., Y. A. Hegazy, A. H. Elkabbany, H. Mohammad, A. Ghiaty, T. M. Abdelghany et al. 2018. Biphenylthiazole antibiotics with an oxadiazole linker: an approach to improve physicochemical properties and oral bioavailability. Eur. J. Med. Chem. 143: 1448–1456.

Hannoun, M. H., M. Hagras, A. Kotb, A-A. M. M. El-Attar and H. S. Abulkhair. 2020. Synthesis and antibacterial evaluation of a novel library of 2-(thiazol-5-yl)-1,3,4-oxadiazole derivatives against methicillin-resistant *Staphylococcus aureus* (MRSA). Bioorg. Chem. 94: 103364.

Huang, Y., J. N. Alumasa, L. T. Callaghan, R. S. Baugh, C. D. Rae, K. C. Keiler et al. 2019. A small-molecule inhibitor of trans-translation synergistically interacts with cathelicidin antimicrobial peptides to impair survival of *Staphylococcus aureus*. Antimicrob. Agents Chemother. 63(4): 02362–02318.

Janardhanan, J., J. E. Meisel, D. Ding, V. A. Schroeder, W. R. Wolter, S. Mobashery et al. 2016. *In vitro* and *in vivo* synergy of the oxadiazole class of antibacterials with b-lactams. Antimicrob. Agents Chemother. 60(9): 5581–5588.

Kaka, A. S., A. M. Rueda, S. A. Shelburne III, K. Hulten, R. J. Hamill and D. M. Musher. 2006. Bactericidal activity of orally available agents against methicillin-resistant *Staphylococcus aureus*. J. Antimicrob. Chemother. 58(3): 680–683

Kim, A., N. M. Wolf, T. Zhu, M. E. Johnson, J. Deng, J. L. Cook et al. 2015. Identification of Bacillus anthracis PurE inhibitors with antimicrobial activity. Bioorg. Med. Chem. 23(7): 1492–1499.

Kotb, A., N. S. Abutaleb, M. Hagras, A. Bayoumi, M. M. Moustafa, A. Ghiaty et al. 2019. tert-Butylphenylthiazoles with an oxadiazole linker: a novel orally bioavailable class of antibiotics exhibiting antibiofilm activity. RSC Adv. 9: 6770–6778.

Kwon, H. C., A. P. D. M. Espindola, J.-S. Park, A. Prieto-Davo, M. Rose, P. R. Jensen et al. 2010. Nitropyrrolins, Cytotoxic farnesyl-a-nitropyrroles from a marine-derived bacterium within the actinomycete family Streptomycetaceae. J. Nat. Prod. 73(12): 2047–2052.

Liu, J.-C., C-J. Zheng, M.-X. Wang, Y.-R. Li, L.-X. Ma, S.-P. Hou et al. 2014. Synthesis and evaluation of the antimicrobial activities of 3-((5-phenyl1,3,4-oxadiazol-2-yl)methyl)-2-thioxothiazolidin-4-one derivatives. Eur. J Med. Chem. 74: 405–410.

Liu, W., W. Ou-Yang, C. Zhang, Q. Wang, X. Pan, P. Huang et al. 2020. Synthetic polymeric antibacterial hydrogel for methicillin-resistant *Staphylococcus aureus*-infected wound healing: nano-antimicrobial self-assembly, drug- and cytokine-free strategy. ACS Nano 14(10): 12905–12917.

Liu, X., X. Song, Y. Liu, M. Xie, W. Yu, S. Yan et al. 2018. Novel 5H-[1,2,4] oxadiazolo[4,5-a]pyrimidin-5-one derivatives as antibacterial and anticancer agents: synthesis and biological evaluation. Tetrahedron Lett. 59(42): 3767–3772.

Luqman, A., V. L. Blair, R. Brammananth, P. K. Crellin, R. L. Coppel and P. C. Andrews. 2015. Powerful antibacterial activity of phenyl-thiolatobismuth(III) complexes derived from oxadiazolethiones. Eur. J. Inorg. Chem. 2015(29): 4935–4945.

Manukumar, H. M., B. Chandrasekhar, K. P. Rakesh, A. P. Ananda, M. Nandhini, P. Lalitha et al. 2017. Novel T-C@AgNPs mediated biocidal mechanism against biofilm associated methicillin-resistant Staphylococcus aureus (Bap-MRSA) 090, cytotoxicity and its molecular docking studies. Med. Chem. Comm. 8(12): 2181–2194

Matore, B. W., P. Banjare, T. Guria, P. P. Roy and J. Singh 2022. Oxadiazole derivatives: Histone deacetylase inhibitors in anticancer therapy and drug discovery. Eur. J. Med. Chem. Rep. 5: 100058.

Mohammad, H., A. S. Mayhoub, A. Ghafoor, M. Soofi, R. A. Alajlouni, M. Cushman et al. 2014. Discovery and characterization of potent thiazoles versus methicillin- and vancomycin-resistant *Staphylococcus aureus*. J. Med. Chem. 57(4): 1609–1615.

Mohammad, H., P. V. N. Reddy, D. Monteleone, A. S. Mayhoub, M. Cushman, G. K. Hammac et al. 2015. Antibacterial characterization of novel synthetic thiazole compounds against methicillin-resistant Staphylococcus pseu-dintermedius. PloS One 10: e0130385.

Muraglia, E., S. Altamura, D. Branca, O. Cecchetti, F. Ferrigno, M. V. Orsale et al. 2008. 2-Trifluoroacetylthiophene oxadiazoles as potent and selective class II human histone deacetylase inhibitors. Bioorg. Med. Chem. Lett. 18(23): 6083–6087.

Naclerio, G. A., C. W. Karanja, C. Opoku-Temeng and H. O. Sintim. 2019. Antibacterial small molecules that potently inhibit *Staphylococcus aureus* lipoteichoic acid biosynthesis. Chem. Med. Chem. 14(10): 1000–1004.

Naclerio, G. A., N. S. Abutaleb, K. I. Onyedibe, M. N. Seleem and H. O. Sintim. 2020. Potent trifluoromethoxy, trifluoromethylsulfonyl, trifluoromethylthio and pentafluorosulfanyl containing (1,3,4-oxadiazol-2-yl)benzamides against drug-resistant Gram-positive bacteria. RSC Med. Chem. 11: 102–110.

Narasimhamurthy, K. H., S. Chandrappa, S. Sharath Kumar, K. B. Harsha, H. Ananda and K. S. Rangappa. 2014. Easy access for the synthesis of 2-aryl 2,3-dihydroquinazolin-4(1H)-ones using gem-dibromomethylarenes as synthetic aldehyde equivalent. RSC Adv. 4(65): 34479–34486.

National Center for Health Statistics, National vital statistics report. http:// www.infoplease.com/ipa/A0005148.html.

Opoku-Temeng, C., G. A. Naclerio, H. Mohammad, N. Dayal, N. S. Abutaleb, M. N. Seleem et al. 2018. N-(1,3,4-oxadiazol-2-yl) benzamide analogs, bacteriostatic agents against methicillin- and vancomycin-resistant bacteria. Eur. J. Med. Chem. 155: 797–805.

Palumbo Piccionello, A., R. Musumeci, C. Cocuzza, C. G. Fortuna, A. Guarcello, P. Pierro et al. 2012. Synthesis and preliminary antibacterial evaluation of linezolid-like 1,2,4-oxadiazole derivatives. Eur. J. Med. Chem. 50: 441–448.

Pidugu, V. R., N. S. Yarla, A. Bishayee, A. M. Kalle and A. K. Satya. 2017. Novel histone deacetylase 8-selective inhibitor 1,3,4-oxadiazole-alanine hybrid induces apoptosis in breast cancer cells. Apoptosis 22(11): 1394–1403.

Pidugu, V. R., N. S. Yarla, S. R. Pedada, A. M. Kalle and A. K. Satya. 2016. Design and synthesis of novel HDAC8 inhibitory 2,5-disubstituted-1,3,4-oxadiazoles containing glycine and alanine hybrids with anti-cancer activity. Bioorg. Med. Chem. 24(21): 5611–5617.

Pinzi, L., R. Benedetti, L. Altucci and G. Rastelli. 2020. Design of dual inhibitors of histone deacetylase 6 and heat shock protein 90. ACS Omega 5(20): 11473–11480.

Rajak, H., A. Agarawal, P. Parmar, B. S. Thakur, R. Veerasamy, P. C. Sharma et al. 2011. 2,5-Disubstituted-1,3,4-Oxadiazoles/Thiadiazole as surface recognition moiety: design and synthesis of novel hydroxamic acid based histone deacetylase inhibitors. Bioorg. Med. Chem. Lett. 21(19): 5735–5738.

Raju, R., A. M. Piggott, L. X. Barrientos Diaz, Z. Khalil and A.-C. Capon. 2010. Heronapyrroles, Farnesylated 2-nitropyrroles from an Australian marine-derived Streptomyces sp. Org. Lett. 12(22): 5158–5161.

Rakesh, K. P., H. K. Vivek, H. M. Manukumar, C. S. Shantharam, S. N. A. Bukhari, H.-L. Qin et al. 2018. Promising bactericidal approach of dihydrazone analogues against biofilm forming Gram-negative bacteria and molecular mechanistic studies. RSC Adv. 8(10): 5473–5483.

Rane, R. A., P. Bangalore, S. D. Borhade and P. K. Khandare. 2013. Synthesis and evaluation of novel 4-nitropyrrole-based 1,3,4-oxadiazole derivatives as antimicrobial and anti-tubercular agents. Eur. J. Med. Chem. 70: 49–58.

Ravindar, L., S. N. A. Bukhari, K. P. Rakesh, H. M. Manukumar, H. K. Vivek, N. Mallesha et al. 2018. Aryl fluorosulfate analogues as potent antimicrobial agents: SAR, cytotoxicity and docking studies. Bioorg. Chem. 81: 107–118.

Rohand, T., Y. Ramli, M. Baruah, J. Budka and A. M. Das. 2019. Synthesis, structure elucidation and antimicrobial properties of new bis-1,3,4-oxadiazole derivatives. Pharm. Chem. J. 53: 150–154.

Salassa, G. and A. Terenzi. 2016. Metal complexes of oxadiazole ligands: an overview. Int. J. Mol. Sci. 20(14): 3483.

Santo, R. D., R. Costi, M. Artico, S. Massa, G. Lampis, D. Deidda et al. 1998. Pyrrolnitrin and related pyrroles endowed with antibacterial activities against Mycobacterium tuberculosis. Bioorg. Med. Chem. Lett. 8(20): 2931–2936.

Seleem, M. A., A. M. Disouky, H. Mohammad, T. M. Abdelghany, A. S. Mancy, S. A. Bayoumi et al. 2016. Second-generation phenylthiazole antibiotics with enhanced pharmacokinetic properties. J. Med. Chem. 59(10): 4900–4912.

Singla, N., G. Singh, R. Bhatia, A. Kumar, R. Kaur and S. Kaur. 2020. Design, synthesis and antimicrobial evaluation of 1,3,4-oxadiazole/1,2,4-triazole-substituted thiophenes. Chemistry 5(13): 3835–3842.

Spoering, A. L. and K. Lewis. 2001. Biofilms and planktonic cells of Pseudomonas aeruginosa have similar resistance to killing by antimicrobials. J. Bacteriol. 183(23): 6746–6751.

Tian, C., S. Huang, Z. Xu, W. Liu, D. Li, M. Liu, et al. 2022. Design, synthesis, and biological evaluation of β-carboline 1,3,4-oxadiazole-based hybrids as HDAC inhibitors with potential antitumor effects. Bioorg. Med. Chem. Lett. 64(2022): 128663.

Tranchimand, S., C. M. Starks, I. I. Mathews, S. C. Hockings and T. J. Kappock. 2011. Treponema denticola pure is a bacterial AIR carboxylase. Biochemistry 50(21): 4623–4637.

Valente, S., D. Trisciuoglio, T. De Luca, A. Nebbioso, D. Labella, A. Lenoci et al. 2014. 1,3,4-Oxadiazole containing histone deacetylase inhibitors: anticancer activities in cancer cells. J. Med. Chem. 57(14): 6259–6265.

Verma, S. K., R. Verma, K. S. S. Kumar, L. Banjare, A. B. Shaik, R. R. Bhandare et al. 2021a. A key review on oxadiazole analogs as potential methicillin-resistant *Staphylococcus aureus* (MRSA) activity: Structure-activity relationship studies. Euro. J. Medicinal Chem. 2019: 113442.

Verma, S. K., R. Verma, S. Verma, Y. Vaishnav, S. P. Tiwari and K. P. Rakesh. 2021b. Anti-tuberculosis activity and its structure-activity relationship (SAR) studies of oxadiazole derivatives: a key review. Eur. J. Med. Chem. 209: 112886.

Wang, C. Y., C. W. Chiu, K. Muraoka, P. D. Michie and G. T. Bryan. 1975. Antibacterial activity of nitropyrroles, nitrothiophenes, and aminothiophenes *in vitro*. Antimicrob. Agents Chemother. 8(2): 216–219.

Yang, F., P. Shan, N. Zhao, D. Ge, K. Zhu, C. Jiang et al. 2019a. Development of hydroxamate-based histone deacetylase inhibitors containing 1,2,4-oxadiazole moiety core with antitumor activities. Bioorg. Med. Chem. Lett. 29(1): 15–21.

Yang, Z., M. Shen, M. Tang, W. Zhang, X. Cui, Z. Zhang et al. 2019b. Discovery of 1,2,4-oxadiazole-Containing hydroxamic acid derivatives as histone deacetylase inhibitors potential application in cancer therapy. Eur. J. Med. Chem. 178: 116–130.

Zabiulla, M. J. Nagesh Khadri, A. Bushra Begum, M. K. Sunil and S. A. Khanum. 2020. Synthesis, docking and biological evaluation of thiadiazole and oxadiazole derivatives as antimicrobial and antioxidant agents. Results Chem. 2: 100045.

Zhang, Y., M. Morar and S. E. Ealick. 2008. Structural biology of the purine biosynthetic pathway. Cell. Mol. Life Sci. 65(23): 3699–3724.

Zheng, Z., Q. Liu, W. Kim, N. Tharmalingam, B. B. Fuchs and E. Mylonakis. 2018. Antimicrobial activity of 1,3,4-oxadiazole derivatives against planktonic cells and biofilm of *Staphylococcus aureus*. Future Med. Chem. 10(3): 283–296.

Chapter 8

Oxadiazole in Medicine and Drug Discovery

1. Introduction

Health concerns became the most important clinical issue as they grew daily. For effective treatment, medicinal chemists have recently begun searching for novel drugs. Numerous heterocyclic drugs are used in clinical studies to treat infectious illnesses (Siwach and Verma 2020). Oxadiazoles—heterocyclic compounds with five members that contain one oxygen and two nitrogen heteroatoms—are effective and efficacious. Oxadiazole has four different structural isomers: 1,2,3, 1,2,4, 1,2,5, and 1,3,4 (Figure 1). Due to the inductive effect of the extra heteroatom in the structure, they are a weak base and are a crucial component of many commercially available drugs for various disease treatments (Desai et al. 2022a). The most popular oxadiazole-containing drugs on the market include nesapidil (antihypertensive), raltegravir (anti-HIV), furamizole (antibiotic), zibotentan (anticancer), tiodazosin (alpha-adrenergic receptor antagonist), and oxolamine (cough suppressant). They display a wide range of biological activities, including antimicrobial, antihypertensive, antimitotic, anti-inflammatory, anticancer, or antiproliferative effects by hybridizing with either another heterocyclic motif or by adding various alkyl or aryl substituents (Ladani et al. 2015; Ustabaş et al. 2020; Desai et al. 2022b). Oxadiazole is a pharmacophore for hybrid moieties since it forms hydrogen bonds with biological receptors (De et al. 2019; Ningegowda et al. 2020).

Due to the prevalence of oxygen, sulfur, and nitrogen rings in pharmacological compounds, efforts in medicinal chemistry frequently focus on imitating such structural motifs. Numerous natural products, such as nucleic acids, amino acids, carbohydrates, vitamins, and alkaloids, comprise heterocycles. Since heterocyclic scaffolds make up more than 85% of bioactive chemical compositions, they play a significant role in medicine and drug design and are, therefore, frequently found in commercially available drugs (Mfuh and Larionov 2015). They have essential qualities like electron donor and withdrawer, acceptor, and donor of H bonds, thus taking part in interactions with receptors or target molecules (Strzelecka et al. 2022). They are equipped with unique properties such as aqueous solubility, polarity,

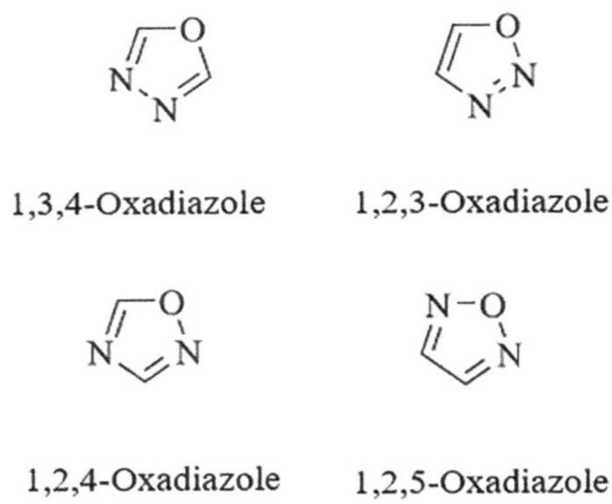

Figure 1. Oxadiazole and their different structural isomers.

lipophilicity, potency through bioisosteric replacements, and selectivity. As a result, they play a much more significant role in modern medicinal chemists' toolkits (Meanwell et al. 2011). Different types of heterocycles, such as pyrimidine, furan, pyrrole, indole, oxadiazole, benzoxazole, azetidine, thiophene, coumarin, benzofuran, pyrazole, quinoline oxazole, quinazoline, pyrimidine, thiazole, quinoline, pyrazole, and isoxazole, exhibits strong pharmacological action (Desai et al. 2022a). For example, pyridine shows a wide range of properties, such as analgesic, anticancer, antimicrobial, antimalarial, antitubercular, anti-inflammatory, antiproliferative, and antiviral effects.

The heterocyclic compounds isolated from nature shared structural characteristics with medicines currently on the market, so minor structural alterations resulted in a variety of heterocyclic molecules that can be used to discover new drugs (Petri et al. 2020; Köprülü et al. 2021). The presence of a variety of functional groups makes them primitive structures of pharmaceutically significant scaffolds (Babaev 1993). Lead optimization is made possible by heterocycles with various shapes and electronic and physicochemical characteristics. However, not all structure-activity relationship (SAR) developments result in the desired pharmacological effect. The ability of heterocycles to participate in hydrogen bonding with the target protein, where the heterocycle can play the role of either H acceptor as in heteroaromatic compounds or H donor as in heterocyclic compounds, can be explained by their improved potency and specificity. Different five- and six-membered heterocycles are built by switching the positions of the aromatic ring's nitrogen, sulfur, and oxygen heteroatoms (Hong et al. 2021). The Food and Drug Administration (FDA) approved numerous drugs in 2020, including remdesivir (broad-spectrum antiviral drug) (Farag et al. 2020), avaprinitib (an antitumor drug) (Rizzo et al. 2021), pematinib (an anticancer drug used to treat cholangiocarcinoma) (Sneyd and Rigby-Jones 2020) remimazolam (sedative drug) (Tan and Habib 2021), and oliceradine (pain medication) (Desai et al. 2018), respectively. In 2021, Asciminib-protein kinase inhibitors and tivozanib-VEGF receptor tyrosine kinase inhibitors were approved, followed by Daridorexant (a medication used to treat insomnia) and Mitapivat (a drug used to treat hemolytic anemia), both were given FDA approval in 2022 (Desai et al. 2022b). The widespread use of the oxadiazole moiety in medicine has led to its emergence

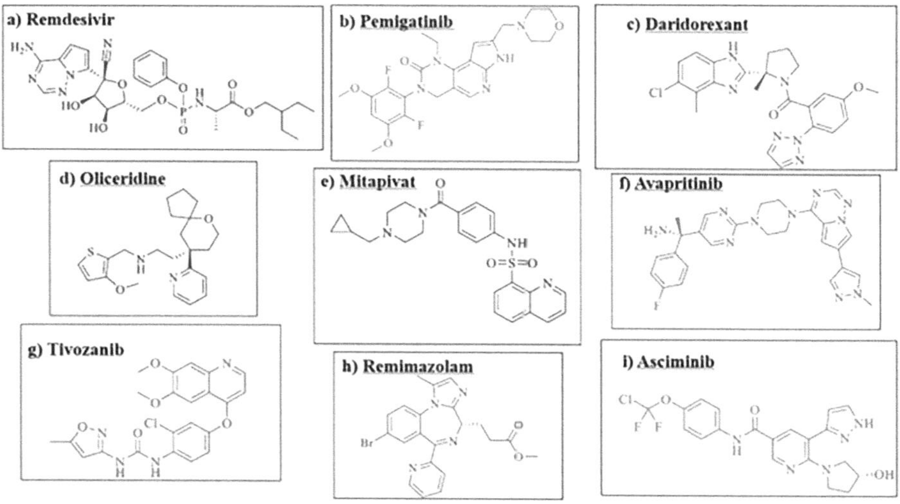

Figure 2. A compilation of FDA approved drugs in recent years (Source: 10.1002/ardp.202200123).

as a separate class of pharmaceutically active compounds, and its aggregation with other heterocyclic scaffolds has produced important bioactive molecules. Figure 2 shows the list of FDA-approved drugs in the recent past.

2. Different Types of Conjugates

2.1 Oxadiazole-Pyridine Conjugate

The antitubercular activity of pyridine conjugates with 1,3,4-oxadiazole derivatives against MTB H37Ra and *M. bovis* Bacillus Calmette-Guérin (BCG) versus RIF and isoniazid as reference drug was evaluated. When MTB H37Ra was in the active or dormant states, the minimum inhibition concentration (MIC) of the derivates was found to be in the range of 1.49–3.43 μg/ml. In order to combat *M. bovis* BCG in active and dormant states, compounds showed MICs in the range of 3.03–19.74 μg/mL, respectively. Additionally, synthetic compounds have a high selectivity index for the human cell line against BCG, which denotes a particular antitubercular profile. A high molecular docking score of between 7.13 and 7.18 for the oxadiazole clubbed pyridine scaffold in the active site of mycobacterial enoyl reductase (InhA) (PDB code: 4TZK) provided essential insights into the binding mode and affinity of the hybrid scaffold. The findings suggested that the scaffold could successfully engage in several close-bonded and nonbonded interactions within the InhA active site (Desai et al. 2018). The triazole and morpholine compounds demonstrated excellent antitubercular activity (25 μg/ml) against MTB H37Rv. The 1,3,4 oxadiazoles are based on 2 benzyl sulfanyl nicotinic acids and have excellent antitubercular activity when tested against MTB H37Rv. These compounds have promising antibacterial properties. The antitubercular activity of these substances was greater than that of the RIF reference drug. Due to the presence of thiazole and benzothiazole units, an excellent antitubercular activity

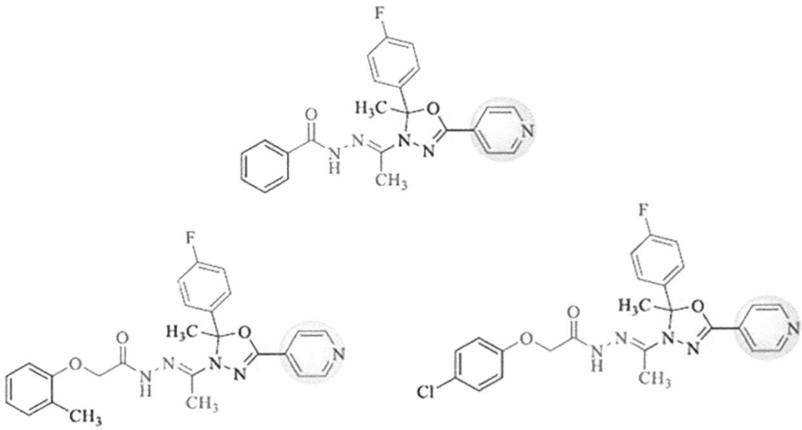

Figure 3. Conjugates of oxadiazole and pyridine exhibit potent antitubercular activity against *M. bovis* (Source: 10.1002/ardp.202200123).

(50–62.5 μg/ml) against MTB H37Rv was determined (Patel et al. 2013). In another study, 1,3,4 oxadiazoles based on pyridine exhibit excellent antitubercular activity against MTB H37Rv at different concentrations of 0.0077, 0.0052, and 0.0089 μM, thereby displaying more than 90% inhibition. Alongside, another derivative of the same conjugate demonstrated antibacterial activity against several Gram-positive and Gram-negative bacteria, including *Escherichia coli* and *Salmonella Paratyphi*, as well as gram-positive bacteria like *Bacillus subtilis* and *Staphylococcus aureus*, respectively (Raval et al. 2014). Oxadiazole-pyridine conjugates with antitubercular activity are presented in Figure 3.

2.2 *Oxadiazole-Benzofuran Conjugates*

A series of 5(5-methyl benzofuran-3-ylmethyl)-3H [1,3,4] oxadiazole-2-thione derivatives demonstrated powerful antitubercular action against MTB H37Rv and *Mycobacterium phlei*. The compounds with chloro and bromo substitutions on benzofuran had outstanding activity against MTB H37Rv and *M. phlei*, with MIC values in the range of 1.56–3.125 μg/ml. The drug's affinity for the MTB MurE ligase has been a significant target for the suppression of TB. The molecular docking studies with a docking score of 109.364 and the compounds' comparable binding poses to the co-crystallized ligand provided additional support for PZA (Negalurmath et al. 2019).

2.3 *Oxadiazole-Thiazole Conjugates*

Compounds developed by Kumar and co-workers (2010) showed MIC of 4 to 8 μg/ml against MTB H37Rv, demonstrating promising antitubercular action. *S. aureus, S. faecalis*, and *B. subtilis*, as well as fungi like *Saccharomyces cerevisiae*, *Candida tropicalis*, and *Aspergillus niger*, were also susceptible to its antimicrobial effects. The 1,3,4 oxadiazoles with pyridine bases were developed by Dhumal et al. (2016) and tested *in vitro* for their antitubercular activity against *M. bovis* BCG and MTB

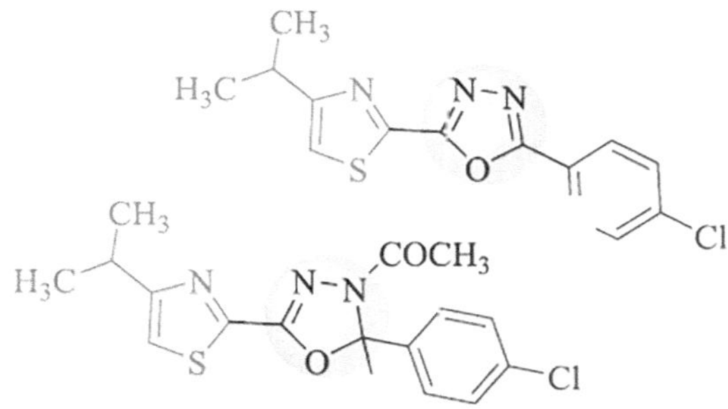

Figure 4. Thiozale-based oxadiazole derivatives with antitubercular activity (Source: 10.1002/ardp.202200123).

H37Ra. The results indicated excellent MIC values between 2.56 and 5.84 μg/ml (Figure 4). The antitubercular activity of theoretical prediction and experimental data were discovered to be connected by the molecular docking investigation. By assessing the degree of affinities to the active site residue, the molecular docking analysis of the produced compounds demonstrated that the compounds were successfully docked into the active site of mycobacterial enoyl reductase. All the active drugs shred similar binding modes in the active site. According to the ligand interaction energy distribution, compounds were found to be anchored to the active site of InhA through several favorable van der Waals interactions, electrostatic interactions, and some stacking interactions. The antitubercular activity of isoxazole clubbed 1,3,4 oxadiazoles was tested against MTB H37Rv by Shingare and co-workers (Shingare et al. 2018). Due to the methoxy group and fluorine atom in the phenyl ring, compounds demonstrated good antitubercular action with MIC of 62.5 and 50 μg/ml. Due to the existence of two aromatic ring replacements on both the isoxazole and oxadiazole rings, the compounds exhibited significant hydrophobic interactions. The MurD ligase enzyme was subjected to molecular docking investigations; the results showed docking scores of 7.28 and 7.35. *In vitro* antitubercular activity against MTB H37Rv (MTB) and isoniazid-resistant MTB (INHR-MTB) strains was evaluated using a series of 1,5-dimethyl-2-phenyl-4-[(5-aryl-1,3,4-oxadiazol-2-yl) methyl] amino-1,2-dihydro-3H pyrazolones (Ahsan et al. 2011). The reported MIC values of 0.78–6.81 and 8.29 μg/ml were 1.4 times more active than isoniazid. The results of the antitubercular activity show that the activity was boosted by the presence of electron-withdrawing groups on the phenyl ring or heterocycle linked to the oxadiazole ring. In a study, 0.78 and 3.12 μg/ml of 1,5-dimethyl-2-phenyl-4-[5-(arylamino)-1,3,4-oxadiazol-2-yl] methylamino-1,2dihydro-3H-pyrazol-3-ones shown potent antitubercular action against MTB and INHR-MTB. Other closely related substances, however, had activity levels of 3.12–6.25 and 6.25–12.5 μg/ml against MTB and INHR-MTB, which was good to moderate. The N-aryl group joined to the oxadiazole ring can be substituted to boost the antitubercular action (Ahsan et al. 2012). With MIC values of 7.46 and 15.01 μg/ml, synthetic 4 (substituted benzylidene) 3 [(5 (pyridine4-yl) 1,3,4oxadiazole2 ylthio] methyl isoxazol5(4H) ones shown strong antitubercular action against MTB H37Ra and *M. bovis* B. Additionally, strong antibacterial activity against Gram-negative strains of *E. coli* and *P. flurescence* (MIC = 2.93 μg/ml and 7.17 μg/ml) was reported, followed by gram-positive bacteria *S. aureus* (MIC = 6.63 μg/ml) and *B. subtilis* (MIC = 3.2 μg/ml) (Chavan et al. 2019). Marine bromopyrrole alkaloids and 1,3,4-oxadiazole were combined using a molecular hybridization method to create hybrids that exhibited broad-spectrum antibacterial action against Gram-positive bacteria *S. aureus* and Gram-negative bacteria *E. coli*. The typical activity ranged from 1.50 to 3.50 μg/ml. The functional groups were changed to increase the antitubercular activity further (Rane et al. 2012). When used against drug-susceptible and drug-resistant TB, a series of N-alkyl-5-(pyridin-4-yl)-1,3,4-oxadiazole-2-amines and related derivatives demonstrated MIC values of 4–8 μM (Vosátka et al. 2018). On the other hand, employing the BACTEC 460 radiometric system, derivatives of 1,3,4 oxadiazole demonstrated greater than 6.25 μg/ml inhibitions against MTB H37Rv in BACTEC 12B medium (Küçükgüzel et al. 2002). A number of Mannich-based compounds were synthesized for *in vitro* antitubercular activity studies against MTB H37Rv and INHR-MTB. The drugs' MIC values against the two antitubercular strains were 0.10–1.10, 0.14–1.14, and 0.42–2.42 μM, demonstrating their excellent antitubercular action (Ali and Shaharyar 2007). The synthesis of 5-aryl-2-thiol,3,4 oxadiazole derivatives and their *in vitro* antitubercular efficacy against MTB H37Rv was reported. At 12.5 μg/ml, the compounds had considerable inhibitory efficacy above 70%. Active substances in Rv1155 were molecularly docked, and it was discovered that these substances fit well in the active site, which is made up of the amino acids Ser144, Arg140, Lys78, Tyr88, Ile74, Leu75, and Leu76. The proteins Ser144 and Arg140 are connected by hydrogen bonds created by the oxygen in the hydroxyl group. The oxadiazole ring was situated in the cavity between the amino acids Leu76 and Leu75, and its nitrogen forms a hydrogen connection with Leu76. The backbone acceptor, backbone donor, and side-chain acceptor characteristics of amino acids were observed (Macaev et al. 2011).

2.4 Piperazine Incorporated 1,3,4-Oxadiazole Derivatives

A group of 1,3,4 oxadiazoles integrating piperazine scaffold were examined for their antibacterial, antitubercular, and antioxidant activities. These compounds had antitubercular activity against MTB with MIC values between 1.60 and 6.25 μg/ml. Three distinct protein targets—isocitratelyase (*Pseudomonas aeruginosa*, PDB ID: 6g1o), dihydrofolate reductase (*S. aureus*, PDB ID: 5isp), and noranthrone synthase—were the subjects of a docking investigation (*Aspergillus parasiticus*, PDB ID: 5kbz). For *in-silico* docking simulations of the antitubercular and antioxidant activity of the produced compounds, methionine aminopeptidase (MTB, PDB ID: 5yxf) and tyrosine-protein kinase (Homo sapiens, PDB ID: 6il3) were chosen, and the compounds have shown remarkable promise. The findings of docking studies indicated that synthetic molecules had higher binding energies to protein targets than conventional medicines. The produced compounds exhibit a stronger affinity for the protein target protein tyrosine kinase 2b (PDB ID: 5tob) in the series and better binding energy and inhibition constant against the protein target enzyme enoyl-acyl carrier protein reductase (PDB ID: 2nsd) (Bhati et al. 2019). As promising antibacterial and antitubercular drugs, 1,3,4 oxadiazole molecules fused with triazolo and tetrazolo groups, and pyrazine was created. The synthesized compounds were evaluated for *in vitro* antitubercular activity against MTB H37Rv, using isoniazid and RIF as reference medications. With a MIC of 6.25 to 50 μg/ml, the compound demonstrated strong antitubercular action against MTB H37Rv (Das et al. 2015). A series of new 1,3,4 oxadiazole-containing quinoline motifs as possible antitubercular drugs. Using the Lowenstein-Jensen medium, the derivatives were tested for MTB H37Rv resistance using RIF and isoniazid as reference medicines. At a 250 μg/ml dosage, four compounds were determined to be the most effective in the main screening, inhibiting MTB H37Rv by more than 90%. The MIC values of the active hybrids ranged from 0.060 to 0.223 mM (Ladani et al. 2015). As possible antitubercular drugs, pyrazine containing 1,3,4 oxadiazoles synthesized and combined with substituted azetidin-2-ones. The synthesized compounds were tested for *in vitro* antitubercular activity against MTB H37Rv using the Lowenstein-Jensen agar method, and isoniazid and RIF were used as reference medications. With a MIC value of 3.12 g/ml, the synthesized compounds were discovered to be the most effective against MTB H37Rv. The antitubercular potential of the compound was moderate (MIC = 6.25 g/ml). High docking scores for compounds were in the range of 69.449 and 64.705 (Das and Mehta 2015). Novel pyrazole-oxadiazole hybrids were synthesized as possible antitubercular drugs, exhibiting MICs of 0.92–2.56 μg/ml against MTB H37Ra in the active state. With MICs between 0.79 and 2.71 μg/ml, compounds showed substantial efficacy against *M. bovis* BCG in the active form. The compounds were successfully docked into the active site of mycobacterial enoyl reductase, according to docking studies (InhA), with a binding energy of 66.459 kcal/mol and a docking score of 10.366 (Desai et al. 2022a). Desai and co-workers synthesized furan and pyridine-based oxadiazole hybrids as possible antitubercular drugs. When tested against MTB H37Ra and *M. bovis* BCG, it showed outstanding antitubercular activity within the range of 97.09 to 98.66% inhibition at a concentration of 30 μg/ml. The molecules were successfully docked into the active site of mycobacterial enoyl reductase, according to a docking study (InhA), with a binding energy of 49.893 kcal/mol and a docking score of 8.991, respectively (Desai et al. 2022a).

2.5 Oxadiazole-Indole, Pyridine, Benzothiazole, and Benzoimidazole Conjugates

Researchers are interested in studying the synthesis and pharmacological properties of pyridine heterocycles because of their biological significance and pharmacological and medical potential. Pyridine has a wide range of biological effects, including anticancer, antimicrobial, anticonvulsant (Prasanthi et al. 2014), antibacterial (Frolova et al. 2011), anti-inflammatory (Brun et al. 2013), and antitumor (Jiao et al. 2015) properties. The nitrogen heterocycle with the pyridine moiety is most likely found throughout nature. Tryptophan, an amino acid, and serotonin, a neurotransmitter, both contain indole moiety as a necessary component. Indole scaffolds display biological activities which are demonstrated by compounds with a 1,3,4-oxadiazole ring (Desai et al. 2018). By using the XTT reduction menadione assay and the nitrate reductase assay, the indole and pyridine-based 1,3,4 oxadiazole compounds against *M. bovis* BCG and MTB H37Ra have been developed and tested for their *in vitro* antitubercular efficacy. RIF and isoniazid were considered as controls. *In vitro* tests of different synthetic compounds against *M. bovis* BCG in both active and dormant states showed potent antitubercular activity. In contrast, other compounds were shown to have excellent antitubercular activity against *M. bovis* BCG in both the active (0.94–5.17 μg/ml) and dormant states (0.85–4.97 μg/ml), respectively. Many other compounds were subjected to a molecular docking study against mycobacterial enoyl reductase (PDB code: 4TZK), which revealed information about the thermodynamic interactions and binding mode with the highest GLIDE docking score, which ranged from 8.07 to 8.267 (Desai et al. 2016). In order to test for *in vitro* antitubercular activity against MTB H37Rv, the benzimidazolyl 1,3,4-oxadiazol 2ylthio-Nphenyl (benzothiazolyl) acetamides were synthesized. The compound showed excellent inhibition activity with a MIC value of 6.25 μg/ml from BACTEC MGIT, a compound with a methoxy group at the para position to the N phenyl acetamide moiety, exhibiting excellent antitubercular activity. The Lowenstein-Jensen MIC method was used to screen these compounds, which displayed MIC values of 12.5 μg/ml, representing their antitubercular potency (Patel et al. 2012).

3. Applications

3.1 Antitubercular Activity

Mycobacterium tuberculosis (MTB), a deadly pathogen that kills more people than any other microbial species, is the cause of tuberculosis (TB). This was first discovered by Robert Koch in 1882. It is the second-leading cause of death, behind COVID-19 (behind HIV/AIDS) (WHO Report 2024). Other microbial species related to MTB, such as *Mycobacterium avium* and *Mycobacterium africanum*, can also cause TB (Beena and Rawat 2013). About 75% of active TB cases are pulmonary TB, while the remaining cases are extrapulmonary TB, in which the bacteria affect the bones, pleura, joints, and lymphatic system (Rivers and Mancera 2008). The World Health Organization (WHO) estimates that 1.5 million people will die worldwide from tuberculosis (TB) in 2020. In the same year, there were 10.0 million new cases of TB worldwide (WHO Report 2021). The first antitubercular drug to be discovered was streptomycin, and its first extensive clinical trial was carried out in 1948, followed by thioacetazone and para-aminosalicylic acid as the potential antitubercular drugs (A Medical Research Council Investigation 1950; Keshavjee and Farmer 2012). High curability rates and decreased susceptibility to antibiotic resistance were produced by combining the two medications (Crofton and Mitchison 1948). However, the widespread emergence of streptomycin drug resistance fueled the creation, research, and development of fresh anti-TB medications (Azevedo and Robinson 2015). These include cycloserine, ethionamide, PZA, rifampicin (RIF), ethambutol, and isonicotinic acid (isoniazid INH). Even though TB is treatable, it still has a high mortality rate and is the most contagious illness in the world. Inadequate case detection remains a significant issue for TB infection control despite the development and expansion of directly observed treatment, short-course programs, and the increased success rate of therapy (Keshavjee and Farmer 2012). Due to the emergence of drug-resistant strains, TB has now become a serious issue. Drug-resistant strains led to the development of both extensive drug-resistant TB (XDR-TB) and multidrug-resistant TB (MDR-TB). While XDR-TB is resistant to any antibiotic, including capreomycin, kanamycin, and amikacin, MDR-TB is resistant to isoniazid and RIF (Caminero and Sotgiu 2010). It is urgently necessary to create hybrid molecules by fusing two or more bioactive heterocyclic moieties into a single molecular scaffold to combat the issue of MDR and TDR TB. To create new hybrids with antitubercular properties, the compounds II, 2 heptadecyl 5 (pyridine 4 yl)1,3,4 oxadiazole, and I, 2 pentadecyl 5 (pyridine 4 yl)1,3,4 oxadiazole, were chosen as two active hybrids against *M. bovis*. These hybrids have antitubercular properties against *M. bovis* with respective minimum inhibitory concentrations (MICs) of 0.35 and 0.65 μM (Navarrete-Vázquez et al. 2007). The emergence of strains that are both multi- and totally drug-resistant has made it necessary to look for more potent molecules to combat microbial species like MTB, which is both multi- and totally drug-resistant.

3.2 Antimalarial Activity

Each year, more than 229 million malaria cases are discovered (WHO Report 2021). More than one hundred countries still experience malaria, which is still a serious health concern in Africa and Asia (Wilkerson et al. 2021). The parasites that caused malaria developed asexually in the host and were spread by female anopheles mosquitoes (Cowman et al. 2016). *Plasmodium falciparum*, along with *Plasmodium falciparum, Plasmodium malariae, Plasmodium vivax, Plasmodium ovale,* and *Plasmodium knowlesi* is the most dangerous species among all the parasites that cause malaria in humans.

The pyrazole acrylic acid-based oxadiazole derivatives were synthesized, which showed *in vitro* antimalarial activity using a culture of the *P. falciparum* chloroquine-sensitive strain. To the chloroquine-sensitive and chloroquine-resistant strains of *P. falciparum*, the compound showed between 0.245 and 0.724 μg/ml. According to molecular docking studies, compounds had docking scores of 5.309 to 6.844, respectively (Verma et al. 2018). Verma and co-workers prepared and tested a series of pyrazole-clubbed oxadiazoles for antimalarial activity. The antimalarial screening results against *P. falciparum,* which is sensitive to chloroquine, indicated the potential for the antimalarial agents with a value of 0.886, 0.248, 0.647, 0.322, 0.582, and 0.494 μg/ml. The catalytic domain of falcipain was shown to interact through hydrogen bonding in the molecular docking study, which resulted in a docking score of 5.532 (Verma et al. 2019). Novel 1,3,4-oxadiazole analogs were synthesized, showing that antibacterial and antimalarial activities *in vitro* are quite effective against *P. falciparum* pathogen. Both pyrimethamine and chloroquine were used as common medications. The potential as a DHFR inhibitor with a binding energy of 7.20 kcal/mol was disclosed by a molecular docking analysis against *P. falciparum* dihydrofolate reductase, which revealed contact with the active site (amino acids) of the receptor (Thakkar et al. 2017). The crystal structure, density functional theory, quantum theory of atoms in molecules, and hybrid quantum mechanics/molecular mechanics binding energy analysis were studied

(Al-Wahaibi et al. 2019) to investigate the potential antimalarial lead candidate 2(4-fluorobenzylthio)5-(5-bromothiophen-2-yl)1,3,4-oxadiazole 95. By blocking the DHFR target with a Glide XP score of less than 5.4 kcal/mol, the molecule may be a viable antimalarial lead candidate, according to the *in-silico* analysis of PfDHFR.

3.3 Anti-Inflammatory Activity

The body's complicated biochemical response to irritants, infections, damaged cells, and other stimuli is inflammation. Inflammation can be acute or persistent. Numerous illnesses, including hay fever, periodontal disease, atherosclerosis, and osteoarthritis, are linked to chronic inflammation. The manufacture of pro-inflammatory prostaglandins from arachidonic acid results in inflammation (Smith et al. 1998; Warner et al. 1999). Biosynthesis involves the cyclooxygenase (COX) enzyme; prostaglandins are produced into pro-inflammatory prostaglandins, which are then released into tissues to cause inflammation. Cyclooxygenase 1 (COX1) and cyclooxygenase 2 (COX2) are the two isoforms of the COX enzyme. The overproduction of prostaglandins during inflammation is caused by COX-2 (Srivastava et al. 2003). Nonsteroidal anti-inflammatory drugs (NSAIDs) relieve pain, fever, and inflammation when used in greater quantities, they also prevent blood clots (Palomer et al. 2002; Merry et al. 2010). Depending on the medication, side effects can include an increased risk of stomach ulcers and bleeding, heart attack, and kidney disease. Derivatives of oxadiazole have intriguing pharmacological characteristics exhibiting strong anti-inflammatory and analgesic effects.

In another study, the compounds with strong anti-inflammatory and antitumor action were synthesized. With ortho, meta, and para substituents in the phenyl ring, a significant reduction in edema was observed, with respective edema inhibition rates of 55%, 67%, and 67% was observed that were very close to those of aspirin (68%) and ibuprofen (73%) (Bezerra et al. 2005). Compounds synthesized by Omar and co-workers (Omar et al. 1996) exhibit strong anti-inflammatory effects on histamine-induced abdominal edema in rats. These compounds had superior anti-inflammatory effectiveness compared to ibuprofen. Based on a 200 mg/kg dose of ibuprofen and mean time (minute standard error), anti-inflammatory actions were considered. Ibuprofen's mean time was 6.33 ± 0.96 min, while compounds 99–103's mean time ranged from 6.91 ± 0.88 to 14.56 ± 0.29 min. The 3-aryl 5-propyl 1, 2, 4 oxadiazole derivatives were synthesized (Srivastava et al. 2003) and evaluated for the development of new NSAIDs that could specifically inhibit the COX2 enzyme, which oversees producing prostaglandins and other mediators linked to the inflammatory process. Using aspirin as a reference medication, the synthesized compounds were tested for their anti-inflammatory efficacy. These substances demonstrated strong anti-inflammatory efficacy, with edema inhibition ranging from 9.73% to 26.57%. Bhandari and co-workers (2008) reported a series of S-substituted phenacyl-1,3,4-oxadiazole-2-thiol derivatives, which demonstrated significant anti-inflammatory activity with percent inhibition values ranging between 71.05% to 74.23% when compared to the reference drug diclofenac (percent inhibition = 71.3%) additionally; these substances had potent analgesic properties. Fenbufen-based 3[5 (substituted aryl)1,3,4oxadiazol2yl]1(biphenyl4yl) propan1ones were claimed to have been synthesized and tested for their analgesic and anti-inflammatory properties. These substances demonstrated considerable anti-inflammatory efficacy, inhibiting edema by more than 55% (Husain et al. 2009). Data on the *in vitro* inhibition of COX enzymes revealed that substances were more selective for COX2 than COX1. The selectivity indices with 4 methoxy and 3, 4 dimethoxy substitutions were 29.9 and 36.1 for COX2, respectively, and their IC_{50} values were 1.5 and 1.8 M. In one study, schiff bases of 1,3,4 oxadiazoles associated quinazolin-4-one with analgesic and anti-inflammatory activities are the best candidate (Dewangan et al. 2017). With 47.84% and 42.33% inhibition, compounds demonstrated more powerful anti-inflammatory action. The development of 1,3,4 oxadiazole-modified benzothiazole derivatives as promising anti-inflammatory and antioxidant compounds is also well documented. Indomethacin served as the reference medication for the *in vivo* para xylene-induced mice ear edema model, which was used to assess the anti-inflammatory effect. At 100 mg/kg, the compound exhibited strong anti-inflammatory action with 57.35% inhibition with IC_{50} values from 0.01 to 0.05 mmol/L, showing strong antioxidant activity (Zheng et al. 2020). Ascorbic acid and diclofenac sodium were used as reference medications for the evaluation of a novel series of 2,5-disubstituted 1,3,4 oxadiazole derivatives for their *in vitro* anti-inflammatory and antioxidant effects. A molecular docking analysis revealed that certain substances might fit snugly inside the COX2 enzyme's active site cavity. The docking score of the compound was 8.063, and its glide energy was 33.293 kcal/mol (Kashid et al. 2020). By using the Griess assay, 3,5-disubstituted-1,3,4-oxadiazole derivatives has been developed as prospective iNOS inhibitors and anti-inflammatory drugs, using indomethacin as the reference medication. In the carrageenan-induced paw edema test, compounds significantly reduced the generation of nitric oxide with concentrations of 12.61 ± 1.16, 11.06 ± 1.34, and 18.95 ± 3.57 μM, at a dose of 100 M. Compounds were shown to be bound in the iNOS active site, according to molecular docking research (Koksal et al. 2021).

3.4 Anti-HIV Activity

Humans can easily contract the spreadable retrovirus, which is a causative agent of AIDS, when an infected person's immune system gradually fails (Hijikata et al. 1998; Baptista et al. 2010). The HIV-1 and HIV-2 viruses are the two varieties out of which HIV-1 spreads more quickly than HIV-2 (Gilbert et al. 2003). The primary cause of HIV infections worldwide is HIV 1. Initial testing is carried out using an enzyme-linked immunosorbent assay to identify HIV-1 antibodies (ELISA). Specimens that tested negative for HIV in the initial ELISA are thought to be HIV-negative. It was discovered that certain oxadiazole hybrids had anti-HIV efficacy. Novel 4 aryl 1, 2, 5 oxadiazole-3 yl carbamate compounds having potential anti-HIV action were synthesized (Takayama et al. 1996). The synthesis of 5-(1-adamantyl)-1,3,4-thiadiazole-2-thiones and 5-(1-adamantyl)-3-substituted aminomethyl-1,3,4-oxadiazoline-2-thiones from 5-(1-adamantyl)-1,3,4-thiadiazoline-2-thione has been synthesized. These compounds inhibited HIV-1 replication in MT Against HIV-1 at 2 μM, the compounds showed moderate anti-HIV activity with more than 10% viral replication (El-Emam et al. 2004). The synthesis of quinolone-based oxadiazoles and screening for their anti-HIV efficacy were reported. Drugs like azidothymidine were commonly utilized. The highest therapeutic indices (TI) values against the HIV-1VB59 and HIV-1UG070 strains were demonstrated. Instead of ortho and meta positions, electron-donating compounds like methyl and methoxy groups on the para position of the phenyl ring improved activity (Shah et al. 2018). In order to create anti-HIV-1 drugs, 1,3,4 oxadiazole-substituted 4-oxo[1,2a] pyrimidine conjugates were synthesized. The anti-HIV activity of compounds was inhibited by 18%, 26%, and 29%, respectively. Drugs like azidothymidine were commonly utilized. The activity was boosted more by substitutions in the para position of the phenyl ring than in the ortho or meta position (Hajimahdi et al. 2013).

3.5 Antimicrobial Activity

Overusing drugs has increased antimicrobial and antibiotic resistance against different bacterial and fungal species (Aminov et al. 2017). Multidrug resistance (MDR) has also contributed to a worrying situation globally (Brown and Wright 2016). Resistance to treatments has increased over the past few decades, necessitating the creation of novel hybrids that exhibit a wide range of antimicrobial properties. In keeping with this, oxadiazoles will be crucial in creating novel antimicrobial drugs. A new series of thiazole clubbed 1,3,4-oxadiazole derivatives were synthesized, and their *in vitro* antibacterial properties were tested. The antibacterial properties of compounds ranged from excellent to good. The substances worked well against *S. aureus*, having corresponding MICs of 500 and 12.5 μg/ml. *E. coli* was susceptible to compounds with corresponding MICs of 50 and 12.5 μg/ml. Some compounds had a 25 μg/ml MIC and were effective against *Candida albicans* (Desai et al. 2013). A series of 3'-acetyl-2'-aryl-5[3'-(6-methylpyridinyl)] compounds like 2,3-dihydro [1,3,4] oxadiazole and 3'-acetyl[2,2'-aryl[3,6-methylpyridinyl]-5'-oxadiazole were synthesized. Evaluation of the antibacterial effects of 2,3-dihydro [1,3,4] oxadiazoles on *S. aureus, E. coli, P. aeruginosa, Aspergillus flavus, Chrysosporium keratinophilum*, and *Candida albicans* was tested. At 1 and 0.5 mg/ml concentrations, compounds exhibit the largest zones of inhibition of 9–13 and 10–13 mm, respectively, demonstrating outstanding antibacterial activity against *S. aureus, E. coli*, and *P. aeruginosa*. The investigated bacterial strains were effectively inhibited by compounds that showed potent antibacterial properties (Shyma et al. 2013). Compared to the common medication fluconazole, compounds showed superior antifungal efficacy against *A. flavus* and *C. keratinophilum*. These chemicals inhibited the enzyme L-glutamine: D-fructose-6-phosphate amidotransferase and successfully docked into the active pocket of GlcN6P (D-fructose-6-phosphate amidotransferase). Compared to typical medications, compounds exhibit lower binding energies toward the target, measuring 3.71, 4.89, 4.27, 3.98, and 4.10 kJ/mol, respectively. Khalilullah et al. synthesized 2 (substituted phenyl) 5 (2,3 dihydro 1,4 benzodioxane 2 yl) 1,3,4 oxadiazoles and tested them for antibacterial activity (Khalilullah et al. 2016). The test compounds were evaluated for antibacterial activity against strains of bacteria from *S. aureus, E. coli*, and *B. subtilis*, as well as for antifungal activity against *A. niger, A. flavus*, and *C. albicans*. Compounds against the tested strains showed significant biological activity. As ketol-acid reductoisomerase (KARI) inhibitors, a new family of Mannich and bis Mannich bases with five substituted 1,3,4-oxadiazoles had been synthesized. The compounds demonstrated outstanding *in vitro* inhibitory effects against rice KARI and good *in vitro* antifungal and herbicidal activity against *Brassica campestris*. Compounds had strong KARI inhibitory actions, with respective Ki values of 3.10 ± 0.71, 0.96 ± 0.42 and 3.86 ± 0.4 µmol/L (Zhang et al. 2016). A number of 5 substituted (1,3,4-oxadiazol-2-yl) quinoline derivatives were synthesized (Salahuddin et al. 2015). Since different Gram-positive strains of *Bacillus cereus* were suppressed at relatively low concentrations of the produced compounds, their *in vitro* antibacterial activity was examined. The derivatives were most effective against *Klebsiella pneumonia* and against *B. cereus* as pure ciprofloxacin. Additionally, it was discovered that their presence significantly reduced the pathogenicity of many Gram-negative bacteria that were multidrug-resistant, such as *Proteus vulgaris* AP169, *Vibro cholera* 765, and *E. coli* 35B. The MIC of the highly active compounds against the tested species was 12.5 µg/ml. As possible antibacterial agents, heterocyclic compounds, including pyrimidine and oxadiazole, were effective against *S. aureus* and *E. coli*, with MIC values of 50 µg/ml and 12.5 µg/ml, respectively. With a MIC value of 50 µg/ml, compounds showed good efficacy against *S. aureus*. With MICs of 100, 50, and 100 µg/ml, respectively, compounds demonstrated promising antifungal efficacy against *C. albicans, A. niger*, and *Aspergillus clavatus*. The results of the docking investigation showed that the compounds were successfully docked into the active site of DNA gyrase (PDB ID: 1KZN) (Desai et al. 2022a). Novel quinoline-fused oxadiazoles showed excellent antibacterial activity with a MIC value of 25–50 µg/ml against *E. coli*. According to the docking investigation, the chemicals were successfully docked into DNA gyrase's active site (PDB ID: 1KZN) (Salahuddin et al. 2015). As possible antibacterial drugs, quinazoline clubbed thiazole and 1,3,4-oxadiazole heterocycles found to be effective against *E. coli* and *S. aureus* with MIC values of 100, 62.5, and 50 µg/ml respectively. The substances were successfully docked into the active site of DNA gyrase, according to a molecular docking analysis (PDB ID: 1KZN) (Desai et al. 2021). As Staphylococcal biofilm inhibitors, 1,2,4-oxadiazole indole analogs were synthesized, which prevented the growth of *S. aureus* ATCC 25923 biofilm and was found to be the most effective. Their respective BIC_{50} values were 9.7, 0.7, and 2.2 µM (Parrino et al. 2021).

3.6 Antitumor/Anticancer/Antiproliferative Activities

According to the WHO, Cancer causes 10 million deaths worldwide, and the top cause of death for both men and women is lung cancer (WHO Report 2020). Chemotherapy can be used to treat cancer when it is first discovered. The success of the treatment depends on receiving timely care, such as a surgical procedure combined with medicines and radiotherapy (Arruebo et al. 2011; Roy and Saikia 2016). When a cell's functioning is out of balance, the cell's gene transcription program results in cell death (Zhao et al. 2022; Siegel et al. 2022). As a result, the body's cells proliferate out of control and spread throughout the entire body. With over a hundred different forms, cancer is a major cause of death in both industrialized and developing nations (Miller et al. 2022).

The present treatment strategies are helpful, but the link between MDR and the toxicity of anticancer drugs has made it necessary to develop new chemotherapeutic drugs. It has been noticed that several 1,3,4-oxadiazole-containing compounds will be exploited as anticancer medicines in the future based on their pharmacological action. The issues with toxicity and MDR may be solved through the creation of new chemotherapeutic drugs. 1,3,4-oxadiazole compounds have been reported as tubulin inhibitors and possible cytotoxic agents. With GI values of 71.56% (growth percentage (GP—28.44%) and 72.68% (GP—27.32%)), the synthesized analogs have the highest levels of cytotoxicity among them at 10 μM. This is a higher value in a single-dose assay than the common medications gefitinib and imatinib. With percent GIs of 102.71%, 95.44%, 92.75%, 92.29%, 91.81%, 89.59%, and 89.20%, the compound was extremely active against MDA-MB-468 (breast cancer), HL-60(TB) (leukemia), NCI-H460 (non-small cell lung cancer), LOXIMVI, UACC-62 (melanoma), SR (leukemia), and M14 (melanoma) cell lines (Ahsan et al. 2018). With percent GIs of 136.36%, 133.09%, 132.40%, 170.76%, 156.66%, and 141.96%, the compound was found to be highly active against UO-31 (renal cancer), OVCAR-4 (ovarian cancer), HOP-92 (non-small cell lung cancer), 786-O (renal cancer), RXF 393 (renal cancer), and SNB-75 (CNS cancer). Out of 55 cancer cell lines, the compound was shown to be effective against 22–35. In common cell lines, compounds show greater cytotoxicity than the established medications gefitinib and imatinib. With GI50 values ranging from 1.61 to > 100 μM, the compound shows considerable cytotoxicity against the cancer cell types under investigation. With respect to the tested cell line SNB75, the compound had the highest GI50 value (1.61 μM). The TGI and LC50 values were respectively 3.36 and > 100 μM. The average cytotoxicity was determined to be 14.38 μM for all examined cancer cell lines. Tubulin polymerization is inhibited *in vitro* with IC_{50} values of 2.8 ± 0.3 and 2.2 ± 0.1 μM, respectively. When compared to the known tubulin inhibitors, compounds covered the pharmacophoric characteristics. The oxadiazoles docked at the active site of tubulin, hence tubulin inhibition accounts for the likely mode of action. 2,4′-bis mercapto oxadiazole diphenylamine derivatives were successfully tested for their potential to inhibit EGFR tyrosine kinase (49% and 50% at 10 μM) and have antiproliferative effects on human breast cancer cells (MCF-7) in the range of 0.73–2.00 μM. According to molecular docking tests, a compound formed a hydrogen bond with Met793, which engage strongly with the back of ATP-binding sites, including those for lapatinib, since they had benzyl and allyl substituents (Abou-Seri et al. 2010). On the other hand, with the benzyl group, an arene-arene interaction with Arg841 was observed. A series of benzimidazole-linked oxadiazole compounds were tested for anticancer efficacy. By having IC_{50} values that were like the common medication gefitinib (0.081 and 0.098 μM, respectively), compounds showed improved binding affinity to the EGFR. With IC_{50} values of 5.0 and 2.55 μM, the compounds were more cytotoxic than 5 fluorouracil against the MCF-7 cancer cell line. The interaction between N1 and Asp831 was demonstrated by molecular docking studies. Due to the same hydrogen bond between Asp831 and N1 of the benzimidazole in both molecules, their binding modes are comparable (Akhtar et al. 2017). Several imidazopyrimidinyl-1,3,4 oxadiazole hybrids were produced and tested against various cancer cell lines, which showed strong anticancer activity with a mean GP of 33.54. At a single dose of 10 µM, it has strong cytotoxic action against most tumor cell lines, with a GI50 value of 1.30–5.46 µM. Additional testing at five different dose levels (0.01, 0.1, 1, 10, and 100 M) revealed GI50 values between 1.30 and 5.64 µM against the examined cancer cell lines. At a single dose of 10 µM, the compound had the strongest activity against HOP-62 (non-small cell lung cancer), with a GP of 65.54% to 72.94% (Subba Rao et al. 2016).

Furthermore, a Topo II-mediated DNA relaxation experiment shows that the same substance can decrease Topo II activity. As demonstrated by molecular modeling and the docking score of 8.29, the Topoisomerase enzyme is bound to it. In a series of novel 5pyridyl1,3,4oxadiazole derivatives, the anticancer and cytotoxic activity was investigated (Khalil et al. 2015). The 2 (benzylsulphanyl)-5aryl-1,3,4-oxadiazole derivatives had EGFR inhibitory action, and the cytotoxic effect of 5(pyridyl)-1,3,4-oxadiazolethiol derivatives against the MCF-7 breast cancer cell line was assessed. Docking studies were conducted in order to determine the likelihood that the target compounds may eventually develop into lead compounds as a result of additional biological research. With IC_{50} values of 0.010 and 0.012 µM and binding energy scores of –10.32 and –10.25, respectively, the compounds were more active than the reference medication. With respective IC_{50} values of 0.036 and 0.037 µM and binding energy scores of–9.98 and–9.79, compounds 187 and 188 were active. A series of 3-(5-substituted-1,3,4-oxadiazol-2-yl) quinolin-2(1H)-oneand3-[5-(2-phenoxymethyl/naphthyloxymethyl-benzoimidazol-1-ylmethyl)-[1,3,4]-oxadiazol-2-yl]–2-p-tolyloxy-quinoline derivatives were synthesized (Salahuddin et al. 2014a). The National Cancer Institute examined these chemicals for anticancer potential and assessed their performance on a panel of 60 cell lines (NCI). The average GP against the cancer cell lines under investigation was determined to be 66.23% for the tested compounds; the GI50 values ranged from 1.41 to 15.8 µM in the one-dose experiment and from 0.40 to 14.9 µM in the five-dose assay. Previously, 2-(naphthalen-1-ylmethyl/naphthalen-2-yloxymethyl)–1-[5-(substitutedphenyl)-[1,3,4]-oxadiazol-2-ylmethyl]–1H-benzimidazoles were synthesized which showed *in vitro* anticancer efficacy against 60 cancer cell lines with a mean GP of 72.85–74.09 against the cancer cell lines under study. The compounds were very active with growth percent of 32.73 and 47.56 for MDA-MB-468 (breast cancer), SK-MEL28 (melanoma), NCI-H522 (non-small cell lung cancer), and UO-31 (renal cancer) (Salahuddin et al. 2014b). Several novel 2,5 disubstituted 1,3,4 oxadiazole derivatives were synthesized, demonstrating a sizable amount of DNA photocleavage activity. According to the current study, oxadiazole derivatives had greater potential for DNA photocleavage than isonicotinoyl hydrazones, and some changes to the basic structure could eventually lead to the development of some chemotherapeutic drugs (Kumar et al. 2015). Furthermore, 1,3,4-oxadiazole compounds were tested for their anticancer and cytotoxic properties which exhibited strong *in vitro* anticancer properties and cytotoxic activity. The percentage viability IC_{50} against hepatocellular carcinoma HepG2 was 21.2 and 12.4 µg/ml, respectively. Compound derivatives showed moderate anticancer activity against breast cancer MCF7 with a percentage inhibition range of 26–50 µg/ml (Bondock et al. 2012). N-benzyl-1-(5-

aryl-1,3,4-oxadiazol-2-yl)−1-(1H-pyrrol-2-yl) methanamines via one-pot reaction of appropriate benzylamine, pyrrole-2-carbaldehyde (N-isocyanimino), triphenyl phosphorane and carboxylic acid were synthesized (Ramazani et al. 2014). Using an MTT assay, the anticancer effects of compounds were evaluated against several cancer cell lines. In comparison to the conventional medication doxorubicin, these substances demonstrated equivalent or superior cytotoxic activity against the A549, HT29, and HT1080 cells with IC_{50} values of 27.5, 20, 25.7, 13.3, 18, 25.1, 17.3, 20, and 45.3 μM, respectively. With IC_{50} values of 13.9 and 4.3 μM, the compound demonstrated superior cytotoxic action against the MCF-7 cancer cell line compared to doxorubicin, which was four times as strong as doxorubicin, a commonly used drug (Desai et al. 2022a).

By using intercalation binding of 1-substituted phenyl 3-(2-oxo-1,3,4-oxadiazol-5-yl)-carbolines, the anticancer efficacy and ctDNA were investigated. With a GI50 in the range of 0.67–3.20 μM against five cancer cell lines—breast (MCF7), resistant ovarian (NCIADR/RES), lung (NCIH460), ovarian (OVCAR03), and colon (HT29) cancer cell lines—a compound with an N, N dimethylaminophenyl group demonstrated the best efficacy with the measured mean of GI50 was 1.68 μM (Savariz et al. 2014). In another study, a new variety of 5H-dibenzo [b,e] azepine-6,11-dione derivatives with 1,3,4-oxadiazole units as possible anticancer drugs for human ovarian cancer cell lines (OVCAR-3) has been produced. Using the MTT assay and rucaparib as a reference medication, the investigated compounds were evaluated against the OVCAR-3 cell line. The most effective compounds against OVCAR-3 were compound 205 and compound 206, with IC_{50} values of 1.66 ± 0.23 and 1.40 ± 0.30 μM, respectively. The strongest molecule, 206, was docked into the PARP-1 active site (PDB ID: 4RV6). Rucaparib's position score was slightly higher than the GBVI/WSA binding free energy of 206 in the S field, which was 13.4189 kcal/mol (He et al. 2018). As possible anticancer medicines, 1,3,4-oxadiazole fused tetrazole amide derivatives have been synthesized. Using the MTT assay and the standard medication doxorubicin, the antitumor efficacy of the compounds was evaluated using three distinct human cancer cell lines A549, MDA-MB-231, and MCF7. Compared to doxorubicin, compounds have the highest potency. With IC_{50} values of 1.77, 2.45, and 1.22 μM for the A549, MDA-MB-231, and MCF-7 cell lines, the compound exhibited strong anticancer activity. With IC_{50} values of 1.02, 1.34, and 0.31 M for the A549, MDA-MB-231, and MCF7 cell lines, the compound exhibited strong anticancer activity. The compound showed promising anticancer activity with IC_{50} values of 1.11, 2.90, and 1.90 μM against the A549, MDA-MB-231, and MCF7 cell lines (Kotla et al. 2020). Also, 2(N-heterocycle) substituted 1,3,4 oxadiazoles were synthesized as possible anticancer drugs. Using the MTT assay and standard medications doxorubicin and 5 fluorouracil, the synthesized compounds were tested for their anticancer efficacy against the three distinct human cancer cell lines MCF7, HT-29, and HepG2. With IC_{50} values of 0.48 ± 0.25, 1.23 ± 0.90, and 3.09 ± 0.49 μM, against the MCF-7, HT-29, and HepG2 cell lines, the compound demonstrated substantial anticancer activity. The compound had IC_{50} values of 0.86 ± 0.47, 0.78 ± 0.19, and 0.26 ± 0.15 against the MCF-7,

HT-29, and HepG2 cell lines. With a binding energy of about 10.9 kcal/mol, the compound was successfully docked into the active site of colon cancer target protein CDK2 (PDB 2R3J) (Bhatt et al. 2020). Novel oxadiazole-based topsentin derivatives were synthesized and demonstrated to have cytotoxic effects on pancreatic cancer cells. With IC_{50} values against Hs766T, PDAC3, HPAF-II, and PATU-T cells ranging from 5.7 to 10.7 μM, the compound was found to be the most effective. Gemcitabine was a commonly prescribed medication. With a docking score of 6.999 kcal/mol, molecular docking demonstrated the potential to connect to the active site of CDK1 (PDB ID: 4YC6) (Pecoraro et al. 2021). In order to develop new topsentin analogs with a 1,2,4-oxadiazole moiety as possible antiproliferative agents against a panel of (NCI-60) cell lines, a new series of compounds were synthesized standard medications like gemcitabine and 5 fluorouracil were utilized. With an EC50 value of 0.40 μM, the compound exhibited strong anticancer activity against SUIT2 cells. The EC50 values for the compound against Panc1 and Capan1 cells were 0.8 and 1.2 μM. With EC50 values of 3.2, 1.6, and 1.3 μM against SUIT2, Panc1, and Capan1 cells, the compound demonstrated antiproliferative efficacy (Carbone et al. 2021). With EC50 values of 2.8 μM, the compound was found to be more effective against Panc1 and Capan1 cancer cell lines. The EC50 values for compound against SUIT2, Panc1, and Capan1 cells were in the range of 2.6 to 6.8 μM. Compounds with EC50 values of 1.5, 1.4, and 1.9 M against Panc1, Capan1, and SUIT2 cells, had equivalent antiproliferative efficacy. The capacity to interact with the ATP-binding sites of GSK-3 was shown by molecular docking (PDB ID: 1UV5).

4. Conclusion and Prospects

The chapter discussed the importance of hybrid heterocyclic compounds bearing oxadiazole moiety and an understanding of their biological targets, showing promising antibacterial, anti-HIV, antitubercular, anticancer, anti-inflammatory, and antimalarial activity. Oxadiazole-containing scaffolds would significantly contribute to the reduction of worldwide mortality from MDR-TB, XDR-TB, and MDR-TB due to significant therapeutic effects. The 1,2,4-, 1,2,5-, and 1,3,4-oxadiazoles and their derivatives appear to provide a viable foundation for therapeutic development. The oxadiazole-based scaffold was found to be promiscuous by molecular docking against a variety of therapeutically important sites and may be exploited to create novel chemotherapeutic drugs in the future. Future development of several new therapeutic candidates, including highly promising derivatives, may be facilitated by the synthetic approach. This will lead to the development of innovative drug design and discovery programs based on oxadiazoles to combat deadly diseases.

References

A Medical Research Council investigation. 1950. Treatment of pulmonary tuberculosis with streptomycin and para-aminosalicylic acid. Br. Med J. 2(4688): 1073–1085.

Abou-Seri, S. M. 2010. Synthesis and biological evaluation of novel 2,4'-bis substituted diphenylamines as anticancer agents and potential epidermal growth factor receptor tyrosine kinase inhibitors. Eur. J. Med. Chem. 45(9): 4113–4121.

Ahsan, M. J., A. Choupra, R. K. Sharma, S. S. Jadav, P. Padmaja, M. Z. Hassan et al. 2018. Rationale design, synthesis, cytotoxicity evaluation, and molecular docking studies of 1,3,4-oxadiazole analogues. Anticancer Agents Med. Chem. 18(1): 121–138.

Ahsan, M. J., J. G. Samy, C. B. Jain, K. R. Dutt, H. Khalilullah and M. S. Nomani. 2012. Discovery of novel antitubercular 1,5-dimethyl-2-phenyl-4-([5-(arylamino)-1,3,4-oxadiazol-2-yl] methylamino)-1,2-dihydro-3H-pyrazol-3-one analogues.Bioorg. Med. Chem. Lett. 22(2): 969–972.

Ahsan, M. J., J. G. Samy, H. Khalilullah, M. S. Nomani, P. Saraswat, R. Gaur et al. 2011. Molecular properties prediction and synthesis of novel 1,3,4-oxadiazole analogues as potent antimicrobial and antitubercular agents. Bioorg. Med. Chem. Lett. 21(24): 7246–7250.

Akhtar, M. J., A. A. Siddiqui, A. A. Khan, Z. Ali, R. P. Dewangan, S. Pasha et al. 2017. Design, synthesis, docking and QSAR study of substituted benzimidazole-linked oxadiazole as cytotoxic agents, EGFR and erbB2 receptor inhibitors. Eur. J. Med. Chem. 126: 853–869.

Akhtar, W., L. M. Nainwal, M. F. Khan, G. Verma, G. Chashoo, A. Bakht et al. 2020. Synthesis, COX-2 inhibition and metabolic stability studies of 6-(4-fluorophenyl)-pyrimidine-5-carbonitrile derivatives as anticancer and anti-inflammatory agents. J. Fluor. Chem. 236: 109579.

Ali, M. A. and M. Shaharyar. 2007. Oxadiazole Mannich bases: synthesis and antimycobacterial activity. Bioorg. Med. Chem. Lett. 17(12): 3314–3316.

Al-Wahaibi, L. H., N. Santhosh Kumar, A. A. El-Emam, N. S. Venkataramanan, H. A. Ghabbour, A. M. S. Al-Tamimi et al. 2019. Investigation of potential anti-malarial lead candidate 2-(4-fluorobenzylthio)-5-(5-bromothiophen-2-yl)-1,3,4-oxadiazole: Insights from crystal structure, DFT, QTAIM and hybrid QM/MM binding energy analysis. J. Mol. Struct. 2019: 230–240.

Aminov, R. 2017. History of antimicrobial drug discovery: Major classes and health impact. Biochem. Pharmacol. 133(1): 4–9.

Arruebo, M., N. Vilaboa, B. Sáez-Gutierrez, J. Lambea, A. Tres, M. Valladares et al. 2011. Assessment of the evolution of cancer treatment therapies. Cancers 3(3): 32793330.

Arshad, M. 2020. Heterocyclic compounds bearing pyrimidine, oxazole and pyrazole moieties: design, computational, synthesis, characterization, antibacterial and molecular docking screening. SN Appl. Sci. 2: 467.

Azevedo, K. J. and T. N. Robinson. 2015. Anthropology in the design of preventive behavioral health programs for children and families living in disadvantaged neighborhoods. Ann. Anthropol. Pract. 39(2): 176–191.

Babaev, E. V. 1993. Molecular design of heterocycles. 2. "Structure-synthesis" magic rule in the synthesis of six-membered heteroaromatic rings (review). Chem. Heterocycl. Compd. 29: 796–817.

Baptista, M. and J. Ramalho-Santos. 2010. Spermicides, microbicides and antiviral agents: recent advances in the development of novel multi-functional compounds. Mini-Rev. Med. Chem. 9(13): 1556–1567.

Beena and D. S. Rawat. 2013. Antituberculosis drug research: a critical overview. Med. Res. Rev. 33(4): 693–764.

Bezerra, N. M. M., S. P. De Oliveira, R. M. Srivastava and J. R. Da. 2005. Synthesis of 3-aryl-5-decapentyl-1,2,4-oxadiazoles possessing antiinflammatory and antitumor properties. Farmaco 60(11-12): 955–960.

Bhandari, S. V., K. G. Bothara, M. K. Raut, A. A. Patil, A. P. Sarkate and V. J. Mokale. 2008. Design, synthesis and evaluation of antiinflammatory, analgesic and ulcerogenicity studies of novel S-substituted phenacyl-1,3,4-oxadiazole-2-thiol and Schiff bases of diclofenac acid as nonulcerogenic derivatives. Bioorg. Med. Chem. 16(4): 1822–1831.

Bhati, S., V. Kumar, S. Singh and J. Singh. 2019. Synthesis, biological activities and docking studies of piperazine incorporated 1, 3, 4-oxadiazole derivatives. J. Mol. Struct. 1191(5): 197–205.

Bhatt, P., A. Sen and A. Jha. 2020. Design and ultrasound-assisted synthesis of novel 1,3,4-oxadiazole drugs for anti-cancer activity. Chem. Select 5(11): 3347–3354.

Bondock, S., S. Adel, H. A. Etman and F. A. Badria. 2012. Synthesis and antitumor evaluation of some new 1,3,4-oxadiazole-based heterocycles. Eur. J. Med. Chem. 48: 192–199.

Brown, E. D. and G. D. Wright. 2016. Antibacterial drug discovery in the resistance era. Nature 529: 336–343.

Brun, P., A. Dean, V. Di Marco, P. Surajit, I. Castagliuolo, D. Carta et al. 2013. Peroxisome proliferator-activated receptor-γ mediates the anti-inflammatory effect of 3-hydroxy-4-pyridinecarboxylic acid derivatives: Synthesis and biological evaluation. Eur. J. Med. Chem. 62: 486–497.

Caminero, J. A., G. Sotgiu, A. Zumla and G. B. Migliori. 2010. Best drug treatment for multidrug-resistant and extensively drug-resistant tuberculosis. Lancet Infect. Dis. 10(9): 621–629.

Carbone, D., B. Parrino, S. Cascioferro, C. Pecoraro, E. Giovannetti, V. Di Sarno et al. 2021. 1,2,4-oxadiazole topsentin analogs with antiproliferative activity against pancreatic cancer cells, targeting GSK3β kinase. Chem. Med. Chem. 16(3): 537–554

Chavan, A. P., R. R. Deshpande, N. A. Borade, A. Shinde, P. C. Mhaske, D. Sarkar et al. 2019. Synthesis of new 1,3,4-oxadiazole and benzothiazolylthioether derivatives of 4-arylmethylidene-3-substituted-isoxazol-5(4H)-one as potential antimycobacterial agents. Med. Chem. Res. 28: 1873–1884.

Cowman, A. F., J. Healer, D. Marapana and K. Marsh. 2016. Malaria: Biology and disease. Cell 167(3): 610–624.

Crofton, J. and D. A. Mitchison. 1948. Streptomycin resistance in pulmonary tuberculosis. Br. Med. J. 2(4588): 1009–1015.

Das, R. and D. K. Mehta. 2021. Evaluation and docking study of pyrazine containing 1, 3, 4-oxadiazoles clubbed with substituted azetidin-2-one: a new class of potential antimicrobial and antitubercular. Drug Res. 71(1): 26–35.

Das, R., G. S. Asthana, K. A. Suri, D. K. Mehta and S. Asthana. 2015. Synthesis and assessment of antitubercular and antimicrobial activity of some novel triazolo and tetrazolo-fused 1, 3, 4-oxadiazole molecules containing pyrazine moiety. J. Pharm. Sci. Res. 7: 806.

De, S. S., M. P. Khambete and M. S. Degani. 2019. Oxadiazole scaffolds in anti-tuberculosis drug discovery. Bioorg. Med. Chem. Lett. 29(16): 1999–2007.

Desai, N. C., K. Bhatt, D. J. Jadeja, H. K. Mehta, V. M. Khedkar and D. Sarkar. 2022b. Conventional and microwave-assisted organic synthesis of novel antimycobacterial agents bearing furan and pyridine hybrids. Drug Dev. Res. 83(2): 416–431.

Desai, N. C., N. Bhatt, H. Somani and A. Trivedi. 2013. Synthesis, antimicrobial and cytotoxic activities of some novel thiazole clubbed 1,3,4-oxadiazoles. Eur. J. Med. Chem. 67: 54–59.

Desai, N., J. Monapara, A. Jethawa, V. Khedkar and B. Shingate. 2022a. Oxadiazole: A highly versatile scaffold in drug discovery. Arch. Pharm. 355(9): e2200123.

Desai, N., N. Shihory, A. Khasiya, U. Pandit and V. Khedkar. 2021. Quinazoline clubbed thiazole and 1,3,4-oxadiazole heterocycles: synthesis, characterization, antibacterial evaluation, and molecular docking studies. Phosphorus, Sulfur Silicon Relat Elem. 196(6): 569–577.

Desai, N. C., A. Trivedi, H. Somani, K. A. Jadeja, D. Vaja, L. Nawale et al. 2018. Synthesis, biological evaluation, and molecular docking study of pyridine clubbed 1,3,4-oxadiazoles as potential antitubercular. Synth. Commun. 48(5): 524–540.

Desai, N. C., H. Somani, A. Trivedi, K. Bhatt, L. Nawale, V. M. Khedkar et al. 2016. Synthesis, biological evaluation and molecular docking study of some novel indole and pyridine-based 1,3,4-oxadiazole derivatives as potential antitubercular agents. Bioorg. Med. Chem. Lett. 26(7): 1776–1783.

Dewangan, D., K. T. Nakhate, V. S. Verma, K. Nagori and D. K. Tripathi. 2017. Synthesis, characterization, and screening for analgesic and anti-inflammatory activities of Schiff bases of 1,3,4-oxadiazoles linked with quinazolin-4-one. J. Heterocycl. Chem. 54(6): 3187–3194.

Dhumal, S. T., A. R. Deshmukh, M. R. Bhosle, V. M. Khedkar, L. U. Nawale, D. Sarkar et al. 2016. Synthesis and antitubercular activity of new 1,3,4-oxadiazoles bearing pyridyl and thiazolyl scaffolds. Bioorg. Med. Chem. Lett. 26(15): 3646–3651.

El-Emam, A. A., O. A. Al-Deeb, M. Al-Omar and J. Lehmann. 2004. Synthesis, antimicrobial, and anti-HIV-1 activity of certain 5-(1-adamantyl)-2-substituted thio-1,3,4-oxadiazoles and 5-(1-adamantyl)-3-substituted aminomethyl-1,3,4-oxadiazoline-2-thiones. Bioorg. Med. Chem. 12(19): 5107–5113.

Farag, S., M. J. Smith, N. Fotiadis, A. Constantinidou and R. L. Jones. 2020. Revolutions in treatment options in gastrointestinal stromal tumours (GISTs): the latest updates. Curr. Treat. Options Oncol. 21(7): 55.

Frolova, L. V., I. Malik, P. Y. Uglinskii, S. Rogelj, A. Kornienko and I. V. Magedov. 2011. Multicomponent synthesis of 2,3-dihydrochromeno[4,3-d]pyrazolo[3,4-b]pyridine-1,6-diones: a novel heterocyclic scaffold with antibacterial activity. Tetrahedron Lett. 52(49): 6643–6645.

Gilbert, P. B., I. W. McKeague, G. Eisen, C. Mullins, A. Guéye-NDiaye, S. Mboup et al. 2003. Comparison of HIV-1 and HIV-2 infectivity from a prospective cohort study in Senegal. Stat. Med. 22(4): 573–593.

Hajimahdi, Z., A. Zarghi, R. Zabihollahi and M. R. Aghasadeghi. 2013. Synthesis, biological evaluation, and molecular modeling studies of new 1,3,4-oxadiazole. Med. Chem. Res. 22(5): 2467–2475.

He, X., X. Li, J. Liang, C. Cao, S. Li, T. Zhang et al. 2018. Design, synthesis and anticancer activities evaluation of novel 5H-dibenzo[b,e]azepine-6,11-dione derivatives containing 1,3,4-oxadiazole units. Bioorg. Med. Chem. Lett. 28(5): 847–852.

Hijikata, M., Y. Ohta, K. Baba, K. Iwata, M. Matsumoto, S. Mishiro et al. 1998. Instability of the NS5A ISDR of hepatitis C virus during natural course: take-over of wild type by mutant type or vice-versa driven by immune pressure. Hepatol. Res. 11(1): 19–25.

Hong, Y. N., J. Xu, G. B. K. Sasa, K. X. Zhou and X. F. Ding. 2021. Remdesivir as a broad-spectrum antiviral drug against COVID-19. Eur. Rev. Med. Pharmacol. Sci. 25(1): 541–548

Husain, A., A. Ahmad, M. M. Alam, M. Ajmal and P. Ahuja. 2009. Fenbufen based 3-[5-(substituted aryl)-1,3,4-oxadiazol-2-yl]-1-(biphenyl-4-yl)propan-1-ones as safer antiinflammatory and analgesic agents. Eur. J. Med. Chem. 44(9): 3798–3804.

Iyer, V. B., B. M. Gurupadayya, B. Inturi and G. V. Pujar. 2016. Synthesis of 1, 3, 4-oxadiazoles as promising anticoagulant agents. RSC Adv. 6(29): 24797–24807.

Jiao, Y., B. T. Xin, Y. Zhang, J. Wu, X. Lu, Y. Zheng et al. 2015. Design, synthesis and evaluation of novel 2-(1H-imidazol-2-yl) pyridine Sorafenib derivatives as potential BRAF inhibitors and anti-tumor agents. Eur. J. Med. Chem. 90: 170–183.

Kashid, B. B., P. H. Salunkhe, B. B. Dongare, K. R. More, V. M. Khedkar and A. A. Ghanwat. 2020. Synthesis of novel of 2, 5-disubstituted 1, 3, 4-oxadiazole derivatives and their *in vitro* anti-inflammatory, anti-oxidant evaluation, and molecular docking study. Bioorg. Med. Chem. Lett. 30(12): 127136.

Kenchappa, R. and Y. D. Bodke. 2020. Synthesis, analgesic and antiinflammatory activity of benzofuran pyrazole heterocycles. Chem. Data Collec. 28: 100453.

Keshavjee, S. and P. E. Farmer. 2012. Tuberculosis, drug resistance, and the history of modern medicine. New Engl. J. Med. 367: 931–936.

Khalil, N. A., A. M. Kamal and S. H. Emam. 2015. Design, synthesis, and antitumor activity of novel 5-pyridyl-1,3,4-oxadiazole derivatives against the breast cancer cell line MCF-7. Biol. Pharm. Bull. 38(5): 763–773.

Khalilullah, H., S. Khan, M. S. Nomani and B. Ahmed. 2016. Synthesis, characterization and antimicrobial activity of benzodioxane ring containing 1,3,4-oxadiazole derivatives. Arabian J. Chem. 9(2): S1029–S1035.

Koksal, M., A. Dedeoglu-Erdogan, M. Bader, E. E. Gurdal, W. Sippl, R. Reis et al. 2021. Design, synthesis, and molecular docking of novel 3,5-disubstituted-1,3,4-oxadiazole derivatives as iNOS inhibitors. Arch. Pharm. 354(8): e2000469.

Köprülü, T. K., S. Ökten, V. E. Atalay, S. Tekin and O. Çakmak. 2021. Biological activity and molecular docking studies of some new quinolines as potent anticancer agents. Med. Oncol. 38: 84.

Kotla, R., A. C. Murugulla, R. Ruddarraju, M. V. B. Rao, P. Aparna and S. Donthabakthuni. 2020. Synthesis and biological evaluation of 1,3,4-oxadiazole fused tetrazole amide derivatives as anticancer agents. Chem. Data Collect. 30: 100548.

Küçükgüzel, S. G., E. F. Oruç, S. Rollas, F. Şahin and A. Özbek. 2002. Synthesis, characterisation and biological activity of novel 4-thiazolidinones, 1,3,4-oxadiazoles and some related compounds. Eur. J. Med. Chem. 37(3): 197–206.

Kumar, G. V. S., Y. Rajendraprasad, B. P. Mallikarjuna, S. M. Chandrashekar and C. Kistayya. 2010. Synthesis of some novel 2-substituted-5-[isopropylthiazole] clubbed 1,2,4-triazole and 1,3,4-oxadiazoles as potential antimicrobial and antitubercular agents. Eur. J. Med. Chem. 45(5): 2063–2074.

Kumar, M., V. Kumar and V. Beniwal. 2015. Synthesis of some pyrazolylaldehyde N-isonicotinoyl hydrazones and 5-disubstituted, 4-oxadiazoles as DNA photocleaving agents. Med. Chem. Res. 24: 2862–2870.

Ladani, G. G. and M. P. Patel. 2015. Novel 1,3,4-oxadiazole motifs bearing a quinoline nucleus: synthesis, characterization and biological evaluation of their antimicrobial, antitubercular, antimalarial and cytotoxic activities. New J. Chem. 39(12): 9848–9857.

Macaev, F., Z. Ribkovskaia, S. Pogrebnoi, V. Boldescu, G. Rusu, N. Shvets et al. 2011. The structure–antituberculosis activity relationships study in a series of 5-aryl-2-thio-1,3,4-oxadiazole derivatives. Bioorg. Med. Chem. 19(22): 6792–6807.

Mazumder, S. A., M. Shahar Yar, R. Mazumder, G. S. Chakraborthy, N. J. Ahsan and M. Ur Rahman. 2017. Updates on synthesis and biological activities of 1,3,4-oxadiazole: A review. Synth. Commun. 47: 1805–1847.

Mbaba, M., L. M. K. Dingle, A. I. Zulu, D. Laming, T. Swart, J. A. De la Mare et al. 2021. Coumarin-annulated ferrocenyl 1,3-oxazine derivatives possessing in vitro antimalarial and antitrypanosomal potency. Molecules 26(3): 1333.

Meanwell, N. A. 2011. Synopsis of some recent tactical application of bioisosteres in drug design. J. Med. Chem. 54: 2529–2591.

Merry, A. F., R. D. Gibbs, J. Edwards, G. S. Ting, C. Frampton, E. Davies et al. 2010. Combined acetaminophen and ibuprofen for pain relief after oral surgery in adults: a randomized controlled trial. Br. J. Anaesth. 104(1): 80–88.

Mfuh, A. M. and O. V. Larionov. 2015. Heterocyclic N-oxides-An emerging class of therapeutic agents. Curr. Med. Chem. 22(24): 2819–2857.

Miller, K. D., L. Nogueira, T. Devasia, A. B. Mariotto, K. R. Yabroff, A. Jemal et al. 2022. Cancer treatment and survivorship statistics. CA Cancer J. Clinc. 72(5): 409–436.

Navarrete-Vázquez, G., G. M. Molina-Salinas, Z. Duarte-Fajardo, J. Vargas-Villarreal, S. Estrada-Soto, F. González-Salazar et al. 2007. Synthesis and antimycobacterial activity of 4-(5-substituted-1,3,4-oxadiazol-2-yl) pyridines. Bioorg. Med. Chem. 15(16): 5502–5508.

Negalurmath, V. S., S. K. Boda, O. Kotresh, P. V. Anantha Lakshmi and M. Basanagouda. 2019. Benzofuran-oxadiazole hybrids: Design, synthesis, antitubercular activity and molecular docking studies. Chem. Data Collect. 19: 100178.

Ningegowda, R., S. Chandrashekharappa, V. Singh, V. Mohanlall and K. N. Venugopala. 2020. Design, synthesis and characterization of novel 2-(2, 3-dichlorophenyl)-5-aryl-1,3,4-oxadiazole derivatives for their anti-tubercular activity against Mycobacterium tuberculosis. Chem. Data Collect. 28: 100431.

Ökten, S., A. Aydın, U. M. Koçyiğit, O. Çakmak, S. Erkan, C. A. Andac et al. 2020. Quinoline-based promising anticancer and antibacterial agents, and some metabolic enzyme inhibitors. Arch. Pharm. 353(9): 2000086.

Omar, F. A., N. M. Mahfouz and M. A. Rahman. 1996. Design, synthesis and antiinflammatory activity of some 1,3,4-oxadiazole derivatives. Eur. J. Med. Chem. 31(10): 819–825.

Palomer, A., F. Cabré, J. Pascual, J. Campos, M. A. Trujillo, A. Entrena et al. 2002. Identification of novel cyclooxygenase-2 selective inhibitors using pharmacophore models. J. Med. Chem. 45(7): 1402–1411.

Parrino, B., D. Carbone, S. Cascioferro, C. Pecoraro, E. Giovannetti, D. Deng et al. 2021. 2,4-Oxadiazole topsentin analogs as staphylococcal biofilm inhibitors targeting the bacterial transpeptidase sortase A. Eur. J. Med. Chem. 209: 112892.

Parrino, B., D. Carbone, S. Cascioferro, C. Pecoraro, E. Giovannetti, D. Deng et al. 2021. 1,2,4-Oxadiazole topsentin analogs as staphylococcal biofilm inhibitors targeting the bacterial transpeptidase sortase A. Eur. J. Med. Chem. 209: 112892.

Patel, N. B., A. C. Purohit, D. P. Rajani, R. Moo-Puc and G. Rivera. 2013. New 2-benzylsulfanyl-nicotinic acid based 1,3,4-oxadiazoles: their synthesis and biological evaluation. Eur. J. Med. Chem. 62: 677–687.

Patel, R. V., P. K. Patel, P. Kumari, D. P. Rajani and K. H. Chikhalia. 2012. Synthesis of benzimidazolyl-1,3,4-oxadiazol-2ylthio-N-phenyl (benzothiazolyl) acetamides as antibacterial, antifungal and antituberculosis agents. Eur. J. Med. Chem. 53: 41–51.

Pecoraro, C., B. Parrino, S. Cascioferro, A. Puerta, A. Avan, G. J. Peters et al. 2021. A new oxadiazole-based topsentin derivative modulates cyclin-dependent kinase 1 expression and exerts cytotoxic effects on pancreatic cancer cells. Molecules 27(1): 19.

Petri, G. L. I., C. Pecoraro, O. Randazzo, S. Zoppi, S. M. Cascioferro, B. Parrino et al. 2020. New imidazo[2,1-b][1,3,4]thiadiazole aerivatives anhibit FAK phosphorylation and potentiate the antiproliferative effects of gemcitabine through modulation of the human equilibrative nucleoside transporter-1 in peritoneal mesothelioma. Antican. Res. 40(9): 4913–4919.

Prasanthi, G., K. V. S. R. G. Prasad and K. Bharathi. 2014. Synthesis, anticonvulsant activity and molecular properties prediction of dialkyl 1-(di(ethoxycarbonyl)methyl)-2,6-dimethyl-4-substituted-1,4-dihydropyridine-3,5-dicarboxylates. Eur. J. Med. Chem. 73: 97–104.

Ramazani, A., M. Khoobi, A. Torkaman, F. Zeinali Nasrabadi, H. Forootanfar, M. Shakibaie et al. 2014. One-pot, four-component synthesis of novel cytotoxic agents 1-(5-aryl-1,3,4-oxadiazol-2-yl)-1-(1H-pyrrol-2-yl)methanamines. Eur. J. Med. Chem. 78: 151–156.

Rane, R. A., S. D. Gutte and N. U. Sahu. 2012. Synthesis and evaluation of novel 1,3,4-oxadiazole derivatives of marine bromopyrrole alkaloids as antimicrobial agent. Bioorg. Med. Chem. Lett. 22(20): 6429–6432.

Raval, J. P., T. N. Akhaja, D. M. Jaspara, K. N. Myangar and N. H. Patel. 2014. Synthesis and *in vitro* antibacterial activity of new oxoethylthio-1,3,4-oxadiazole derivatives. J. Saudi Chem. Soc. 18(2): 101–106.

Rivers, E. C. and R. L. Mancera. 2008. New anti-tuberculosis drugs in clinical trials with novel mechanisms of action. Drug Discov. Today 13(23-24): 1090–1098.

Rizzo, A., A. D. Ricci and G. Brandi. 2021. Pemigatinib: Hot topics behind the first approval of a targeted therapy in cholangiocarcinoma. Cancer Treat. Res. Commun. 27: 100337.

Rowley, J. A., R. C. Reid, E. K. Y. Poon, K. C. Wu, J. Lim, R. J. Lohman et al. 2020. Potent thiophene antagonists of human complement C3a receptor with anti-Inflammatory activity. J. Med. Chem. 63(2): 529–541.

Roy, P. S. and B. J. Saikia. 2016. Cancer and cure: A critical analysis. Indian J. Cancer 53: 441–442.

Salahuddin, A. Mazumder and M. Shaharyar. 2014a. Synthesis, characterization, and *in vitro* anticancer evaluation of novel 2,5-disubstituted 1,3,4-oxadiazole analogue. Bio. Med. Res. Int. 2014: 491492.

Salahuddin, M. Shaharyar, A. Mazumder and M. J. Ahsan, Arabian. 2014b. Synthesis, characterization and anticancer evaluation of 2-(naphthalen-1-ylmethyl/naphthalen-2-yloxymethyl)-1-[5-(substituted phenyl)-[1,3,4]oxadiazol-2-ylmethyl]-1H-benzimidazole. J. Chem. 7(4): 418–424.

Salahuddin, A. Mazumder and M. Shaharyar. 2015. Synthesis, antibacterial and anticancer evaluation of 5-substituted (1,3,4-oxadiazol-2-yl) quinoline. Med. Chem. Res. 24: 2514–2518.

Savariz, F. C., M. A. Foglio, A. L. T. G. Ruiz, W. F. da Costa, M. de Magalhães Silva, J. C. C. Santos et al. 2014. Synthesis and antitumor activity of novel 1-substituted phenyl 3-(2-oxo-1,3,4-oxadiazol-5-yl) β-carbolines and their Mannich bases. Bioorg. Med. Chem. 22(24): 6867–6875.

Sevinçli, Z. S., G. N. Duran, M. Özbil and N. Karalı. 2020. Synthesis, molecular modeling and antiviral activity of novel 5-fluoro-1H-indole-2,3-dione 3-thiosemicarbazones. Bioorg. Chem. 104: 104202.

Shah, P., D. Naik, N. Jariwala, D. Bhadane, S. Kumar, S. Kulkarni et al. 2018. Synthesis of C-2 and C-3 substituted quinolines and their evaluation as anti-HIV-1 agents. Bioorg. Chem. 80: 591–601.

Shingare, R. M., Y. S. Patil, J. N. Sangshetti, R. B. Patil, D. P. Rajani and B. R. Madje. 2018. Synthesis, biological evaluation and docking study of some novel isoxazole clubbed 1,3,4-oxadiazoles derivatives. Med. Chem. Res. 27: 1283–1291.

Shyma, P. C., B. Kalluraya, S. K. Peethambar, S. Telkar and T. Arulmoli. 2013. Synthesis, characterization and molecular docking studies of some new 1,3,4-oxadiazolines bearing 6-methylpyridine moiety for antimicrobial property. Eur. J. Med. Chem. 68: 394–404.

Siegel, R. L., K. D. Miller, H. E. Fuchs and A. Jemal 2022a. Cancer statistics, 2022. Ca-Cancer J. Clin. 72(1): 7–33.

Siwach, A. and P. K. Verma. 2020. Therapeutic potential of oxadiazole or furadiazole containing compounds. BMC Chemistry 14: 70.

Šlachtová, V. and L. Brulíková. 2018. Benzoxazole derivatives as promising antitubercular agents. Chem. Select 3(17): 4653–4662.

Smith, C. J., Y. Zhang, C. M. Koboldt, J. Muhammad, B. S. Zweifel, A. Shaffer et al. 1998. Pharmacological analysis of cyclooxygenase-1 in inflammation. Proc. Natl. Acad. Sci. U. S. A. 95(22): 13313–13318.

Sneyd, J. R. and A. E. Rigby-Jones. 2020. Remimazolam for anaesthesia or sedation. Curr. Opin. Anaesthesiol. 33(4): 506–511.

Srivastava, R. M., A. de Almeida Lima, O. S. Viana, M. J. da Costa Silva, M. T. J. A. Catanho and J. O. F. de Morais. 2003. Antiinflammatory property of 3-aryl-5-(n-propyl)-1,2,4-oxadiazoles and antimicrobial property of 3-aryl-5-(n-propyl)-4,5-dihydro-1,2,4-oxadiazoles: Their syntheses and spectroscopic studies. Bioorg. Med. Chem. 11(8): 1821–1827.

Strzelecka, M., T. Glomb, M. Drąg-Zalesińska, J. Kulbacka, A. Szewczyk, J. Saczko et al. 2022. Synthesis, anticancer activity and molecular docking studies of novel *n*-mannich bases of 1,3,4-oxadiazole based on 4,6-dimethylpyridine scaffold. Int. J. Mol. Sci. 23(19): 11173.

Subba Rao, A. V., M. V. P. S. Vishnu Vardhan, N. V. Subba Reddy, T. Srinivasa Reddy, S. P. Shaik, C. Bagul et al. 2016. Synthesis and biological evaluation of imidazopyridinyl-1,3,4-oxadiazole conjugates as apoptosis inducers and topoisomerase IIα inhibitors. Bioorg. Chem. 69: 7–19.

Takayama, H., S. Shirakawa, M. Kitajima, N. Aimi, K. Yamaguchi, Y. Hanasaki et al. 1996. Utilization of wieland furoxan synthesis for preparation of 4-aryl-1,2,5-oxadiazole-3-yl carbamate derivatives having potent anti-HIV activity. Bioorg. Med. Chem. Lett. 6(16): 1993–1996.

Tan, H. S. and A. S. Habib. 2021. Oliceridine: A novel drug for the management of moderate to severe acute pain—a review of current evidence. J. Pain Res. 14: 969–979.

Thakkar, S. S., P. Thakor, H. Doshi and A. Ray. 2017. 1,2,4-Triazole and 1,3,4-oxadiazole analogues: Synthesis, MO studies, *in silico* molecular docking studies, antimalarial as DHFR inhibitor and antimicrobial activities. Bioorg. Med. Chem. 25(15): 4064–4075.

Ustabaş, R., N. Süleymanoğlu, Y. Ünver and S. Direkel. 2020. 5-(4-Bromobenzyl)-4-(4-(5-phenyl-1,3,4-oxadiazole-2-yl)phenyl)-2,4-dihydro-3H-1,2,4-triazole-3-one: Synthesis, characterization, DFT study and antimicrobial activity. J. Mol. Str. 1214: 128217.

Verma, G., G. Chashoo, A. Ali, M. F. Khan, W. Akhtar, I. Ali et al. 2018. Synthesis of pyrazole acrylic acid-based oxadiazole and amide derivatives as antimalarial and anticancer agents. Bioorg. Chem. 77: 106–124.

Verma, G., M. F. Khan, L. Mohan Nainwal, M. Ishaq, M. Akhter, A. Bakht et al. 2019. Targeting malaria and leishmaniasis: Synthesis and pharmacological evaluation of novel pyrazole-1,3,4-oxadiazole hybrids. Part II. Bioorg. Chem. 89: 102986.

Vosátka, R., M. Krátký, M. Švarcová, J. Janoušek, J. Stolaříková, J. Madacki et al. 2018. New lipophilic isoniazid derivatives and their 1,3,4-oxadiazole analogues: Synthesis, antimycobacterial activity and investigation of their mechanism of action. Eur. J. Med. Chem. 151: 824–835.

Warner, T. D., F. Giuliano, I. Vojnovic, A. Bukasa, J. A. Mitchell and J. R. Vane. 1999. Nonsteroid drug selectivities for cyclo-oxygenase-1 rather than cyclo-oxygenase-2 are associated with human gastrointestinal toxicity: a full *in vitro* analysis. Proc. Natl. Acad. Sci. U. S. A. 96(13): 7563–7568.

Wilkerson, H., G. Maniam, R. E. Dean, T. Bewaji, E. Okotcha and R. Mattamal. 2021. Pediatric coinfection with malaria and epstein-barr virus. Consultant 61(7): e6.

World Health Organization, Global Tuberculosis Report 2021, World Health Organization, Geneva, Switzerland 2021.

World Health Organization, Global Tuberculosis Report 2024, World Health Organization, Geneva, Switzerland 2024. https://www.who.int/news-room/fact-sheets/detail/tuberculosis.

World Health Organization, WHO Report on Cancer: Setting Priorities, Investing Wisely and Providing Care for All, World Health Organization, Geneva. 2020. https://apps.who.int/iris/bitstream/ hand le/10665/330745/9789240001299-eng.pdf.

World Health Organization, World malaria report. 2020. World Health Organization, Geneva 2020. https://www.who.int/ publications/i/item/9789240015791.

Zhang, Y., X. H. Liu, Y. Z. Zhan, L. Y. Zhang, Z. M. Li, Y. H. Li et al. 2016. Synthesis and biological activities of novel 5-substituted-1,3,4-oxadiazole Mannich bases and bis-Mannich bases as ketol-acid reductoisomerase inhibitors. Bioorg. Med. Chem. Lett. 26(19): 4661–4665.

Zhao, M., Z. Zhu, F. Hao, Y. Song, Z. Wang, Y. Ni et al. 2019. The regulatory role of non-coding RNAs on programmed cell death four in inflammation and cancer. Front. Oncol. 9: 919.

Zheng, X., C. Li, M. Cui, Z. Song and X. Bai. 2020. Synthesis, biological evaluation of benzothiazole derivatives bearing a 1,3,4-oxadiazole moiety as potential anti-oxidant and anti-inflammatory agents. Bioorg. Med. Chem. Lett. 30(13): 127237.

Index